INTRODUCTION TO THERMOMECHANICS OF MAGNETIC FLUIDS

INTRODUCTION TO THERMOMECHANICS OF MAGNETIC FLUIDS

V. G. Bashtovoy
B. M. Berkovsky
A. N. Vislovich
Institute for High Temperatures
USSR Academy of Sciences
Moscow

Edited by:

B. M. Berkovsky

English-Edition Editor:

R. E. Rosensweig
Exxon Research and Engineering Company

HEMISPHERE PUBLISHING CORPORATION
A subsidiary of Harper & Row, Publishers, Inc.
Washington New York London

DISTRIBUTION OUTSIDE NORTH AMERICA

SPRINGER-VERLAG
Berlin Heidelberg New York London Paris Tokyo

INTRODUCTION TO THERMOMECHANICS OF MAGNETIC FLUIDS

Originally published by Institut Vysokikh Temperatur AN SSSR as, Vvedenie V Termomekhaniku Magnitnykh Zhidkostey. English Translation by J. Smith.

1 2 3 4 5 6 7 8 9 0 B C B C 8 9 8 7

This book was set in Times Roman by The Sheridan Press. The cover designer was Don Wittig. BookCrafters, Inc. was printer and binder.

Library of Congress Cataloging-in-Publication Data

Bashtovoy, V. G.
Introduction to thermomechanics of magnetic fluids.

Translation of: Vvednie v termomekhaniku magnitnykh zhidkostei.
"Institute for High Temperatures, USSR Academy of Sciences, Korovinskoye Shosse, Moscow"
Bibliography: p.
Includes index.
1. Magnetic fluids—Thermomechanical properties.
I. Berkovsky, B. M. II. Vislovich, A. N. III. Title.
QC766.M36B3713 1987 538′.4 87-11945
ISBN 0-89116-643-2 Hemisphere Publishing Corporation

DISTRIBUTION OUTSIDE NORTH AMERICA:
ISBN 3-540-18210-1 Springer-Verlag Berlin

CONTENTS

FOREWORD

This book contains primary information on the structure and properties of magnetic fluids, a new promising technological material. The simplest mathematical models of the mechanics, thermodynamics and electrodynamics of magnetic fluids are discussed. Special emphasis is made of certain physical concepts which can help the reader study original works.

The book is written by specialists who have made a considerable contribution to the development of the theory and practical application of magnetic fluids in engineering.

As for its contents, level, and form of presentation, the book is intended for a wide range of readers. First, these are specialists in magnetic hydrodynamics, rheology, thermal physics, heat and mass transfer. The book is of interest to specialists in the field of chemical engineering as well as to engineers and technicians interested in the application of magnetic materials. The book will be of use to undergraduates and post-graduates of physicotechnical and mechano-mathematical faculties of the universities and polytechnics.

PREFACE

Until quite recently there were no liquids that may have been magnetized to a value comparable with the magnetization of solid ferromagnetics. The liquids known so far were either dia- or paramagnets with magnetic susceptibility less than 10^{-3}. The situation changed in the mid-60s with the production of stable colloids of solid ferromagnets, namely artificial liquid magnetics. They were referred to as magnetic fluids (MF), also known in literature as ferrofluids and more often than not, magnetizable fluids. Magnetic fluids differ from para- and diamagnetic fluids not only by their much higher sensitivity to a magnetic field, but also by a much slower magnetic relaxation.

Rapid progress has been made over the past 20 years in the production of magnetic fluids. There have been developed stable low-conductivity MFs with initial magnetic susceptibility as high as tens of units and saturation magnetization up to hundreds of kiloampers per meter. The first steps have been made to create conductive MFs. Large-scale physicochemical studies of MFs are now under way which have taken the shape of an individual trend aimed at studying a new aspect in the interaction between an electromagnetic field and a substance.

Multiple and qualitatively diverse effects of the interaction between MFs and an electromagnetic field provide ample opportunities for their practical application. Magnetic fluids are already being employed in seals, dampers, bearings, separators, heat and mass transfer apparatuses and instrumentation. They are now used in ecology and medicine. They are a new promising technological material whose range of applications is permanently expanding. The diversity of studies is worthy of mention. They include physical chemistry of colloids, physics of magnetic phenomena, and mechanics of continuum. High fluidity alongside strong magnetic properties explains the keen interest taken in MFs as a subject of mechanics. About half of all the publications are devoted to the modelling of isothermal MF flows, the analysis of friction and heat transfer mechanisms and effects, convective stability, to the study of sound generation and propagation. It is precisely these trends of studies that have yielded profound scientific and practical results; they underlie the design and development of a majority of instruments and engineering devices. It therefore seems reasonable to publish a book which might be an introduction to modern studies of MFs via the most developed field, namely the thermomechanics of magnetic fluids.

This book contains primary information on the structure, the physical and the simplest mathematical MF models which is necessary to solve the mechanics and convective heat transfer problems. The most important experimental results are discussed. The authors endeavored to present a consistent and understandable consideration of the basic mechanisms of MF and field interactions that are already the developed concepts, as well as to show their effect on MF behavior in particular situations. It is not the aim of this monograph to give an exhaustive coverage of the reported material. The monograph does not concern the methodology of experimental and numerical studies. Nor are the problems of structurization and stratification of particles, specific features in the interaction of MF and the field in the conditions of high-rate shear flows, when affected by high-frequency nonstationary fields that are now under intensive development. The consideration is based on the idea that a MF is a one-component isotropic one-phase continuum with magnetization that is not greatly different from equilibrium magnetization. Two mathematical MF models are used in the work. One of them is based on the assumption of equilibrium MF magnetization in arbitrary dynamic processes. The basic presumptions of this model are formulated by Neuringer and Rosensweig [36]. The phenomena due to magnetization departure from equilibrium are interpreted within the low-frequency approximation suggested by the authors of [40].

The book is intended for a wide range of readers, primarily for specialists in magnetic hydrodynamics, rheology, thermal physics, mass and heat transfer. It may also be found useful by specialists in chemical engineering, by engineers and technicians who are interested in the application of magnetic materials. The book will be beneficial to the undergraduates and postgraduates of physicotechnical and mechanomathematical faculties of the universities and polytechnics.

The editor hopes that despite inevitable shortcomings, specialists in mechanics and thermophysics may find this book a useful introduction to a new field of continuum mechanics, and engineers may get acquainted with the properties of a promising technological material.

B. M. Berkovsky

CHAPTER

ONE

INTRODUCTION

The rapid development of magnetic fluid investigations, now an independent scientific trend, has been greatly promoted by the progress made in the field of mechanics and electrodynamics of continua and the physics of magnetic phenomena elucidated in monographs [1–4]. The bibliography on magnetic fluids (MF) includes more than 1,500 references; among them more than 400 patents and Inventor's Certificates. Information on these studies is considered in surveys [5, 6]. Magnetic fluids are dealt with in monographs [9, 10]. Some survey papers are available [11–15]. The proceedings of meetings, conferences [16–29] and subject handbooks [30–35] discussing the MF fluids are published. The MF studies may be related to one of the four sections, some particulars of which are the subject of the present paper.

1.1 PRODUCTION OF MAGNETIC FLUIDS

A magnetic fluid is a colloidal solution, a stable suspension of solid particles in a liquid. Stability of the suspension is ensured by Brownian motion of particles. Only fine particles are involved in the Brownian motion. Moreover, ferromagnetic particles may attract one another and coalesce to deposit in the gravitational field or a nonuniform magnetic field in aggregations. In order to prevent aggregation, solid ferromagnetic particles are coated with a layer of surfactant which inhibits their coalescence when drawn together to close distance. As a result, the magnetic fluid is a stable colloidal system consisting of rather small (30–100 Å in size) (solid) surfactant-coated ferromagnetic particles dispersed in a liquid carrier.

Depending on the problems facing the investigator or the user, the MF may be obtained on various bases, such as water, hydrocarbons, fluorinated hydrocarbons, mineral and vacuum oils, silicone liquids, as well as on different ferromagnets such as iron, magnetite (Fe_3O_4) and cobalt. In producing a magnetic fluid on a required base, it is most difficult to choose a surfactant and to obtain fine magnetic particles. There are two possible ways of obtaining appropriate particles of colloidal sizes; they may be grown (condensation methods) or may be obtained by grinding coarse-dispersed particles (dispersion methods).

With the use of dispersion methods, particles are ground in ball or vibratory mills with the aid of ultrasound, electric discharges and other physical actions. The first magnetic fluids were obtained as follows. Magnetite was ball-milled for several weeks in the presence of kerosene or oleic acid as surfactants. The fluids produced were rather expensive. Thus, the Ferrofluidics Corporation (USA), founded in the late 1960s and engaged in the production of MFs and their application in technological devices, sold them for $800. to $9000. per 1 kg.

Of all the numerous condensation methods the most widely spread is the method of chemical condensation involving the precipitation of magnetite particles out of an aqueous solution of ferrous and ferric iron salts with the excess of a concentrated alkali solution. The high yield of particles about 100 Å, simplicity and efficiency of the method (the reaction proceeds in a chemical reactor of simple design with heating and agitation), have made it widely used. This method was also favoured because magnetic fluids with magnetite as a dispersed phase possess magnetic properties that are not inferior to fluids with such strong magnetics as iron or cobalt. The colloid ought to be made more stable, and therefore a much thicker "coat" of surfactant should be applied to particles as their magnetization grows. This lowers the maximum possible volume fraction of the magnetic and, hence, the maximum magnetization of the colloid.

The precipitate of colloidal particles should be converted to a carrier liquid. This procedure is realized by peptization. The latter implies the formation on the surface of particles of a layer of surfactant molecules repulsing one another by their elastic ends that are akin to the liquid carrier. As a result, precipitate particle agglomerates fall into separate surfactant-coated particles which are then evenly distributed in the carrier liquid. Peptization is performed by adding the carrier liquid and the surfactant to the magnetic precipitate during heating and agitation of the solution.

The method is simple, adaptable to streamlined manufacture and allows MFs to be produced on various media. It has become widely spread in recent years resulting in a sharp decrease of the MF cost of production which is slightly higher than the cost of their components.

The interest to the production of MFs based on molten metals is growing. Magnetic fluids based on mercury, gallium, and some alloys already exist but their properties such as magnetization, chemical and aggregative stability are still inferior to the best non-conducting MFs. Concentrated conductive mercury-based MFs are obtained which contain iron particle averages of about 25 Å in size. Iron particles were stabilized by introducing tin. Magnetization ~100 kA/m at $2 \cdot 10^{-2}$ Pa · s viscosity and MF stability within 15 to 20 months were achieved using bismuth and lithium additives. In this case, concentration of the magnetic filler was about 3%.

The commercial production of MFs is faced with such problems as the choice of an appropriate raw material and stabilizer, the design of commercial installation and precipitator, evaluation of corrodibility of the materials (solution) used and treatment of harmful effluents. The toxicity of magnetic fluids intended for use in biological medicine and standardization are acute problems to be solved.

1.2 PROPERTIES OF MAGNETIC FLUIDS

Magnetization

Ferromagnetic particles in a MF are monodomain and possess almost-constant magnetic moments. In the absence of a magnetic field their orientations are disordered by the thermal rotational motion, and the magnetic moment of the unit volume (magnetization) is equal to zero. Thus, the representative MFs are magnetizable media rather than liquid magnets. The static magnetization curve shows the Langevin behavior. Deviations from the Langevin curve may be stipulated by different magnetic moments of particles and their dipole-dipole interaction. In variable fields, magnetization is relaxational. Relaxation times are related to the Brownian rotational diffusion of particles in liquid and Neel alternating magnetization.

Internal Friction

Viscosity is another important characteristic of MFs as a liquid magnetic. The MF viscosity is mainly determined by the viscosity of a carrier liquid. The effect of a solid phase obeys the laws typical for suspensions. In MFs, there exists a specific mechanism of internal friction which appears in case of relative rotation of a material element and the field. The intensity of this mechanism is characterized by the rotational viscosity which is not a transport coefficient by nature, although in some cases it behaves as a dynamic viscosity coefficient. The reciprocal viscosity of concentrated MFs is essentially affected by magnetic dipole-dipole interaction of particles. This interaction gives rise to pseudo-plastic rheological properties. In a magnetic field, these properties are enhanced and transverse anisotropy appears.

Electric Properties

Information about electrical conductance, dielectric constant, loss tangent of a dielectric and breakdown voltage is necessary to use MFs in electrical-engineering devices. Besides, electrical measurements are used for MF diagnostics.

For example, an essential excess of electrical conductivity of dielectric-based MFs against the carrier liquid points to a low quality of MFs. This is due to poor washing of colloidal particles off the impurity ions (which appear during electrolytic precipitation of particles) and decrease the aggregative stability of MFs. Because of residual ions, MF electrical conductance is, at best, two–three orders in excess of the electrical conductance of the carrier liquid. The quantity of residual ions per unit volume is in proportion to the concentration of ferromagnetic particles. As the concentration of particles grows, however, the mobility of ions decreases because of a diminishing volume fraction of the liquid phase, with possible free migration of ions. This is responsible for a maximum in the dependence of electrical conductance on the concentration of the ferromagnetic phase at moderate concentrations. MFs are also distinctive for increasing electrical conductance with temperature, for an appreciable growth of di-electric loss at frequences below 10^5 Hz, and for lower breakdown voltage as compared to the carrier medium.

Optical Properties

Although MFs are optically of low transparency, the light can penetrate their thin layer, the thickness of which increases as the ferrophase concentration decreases. The light attenuates due to absorption and Rayleigh scattering. Absorption and scattering are respectively predominant in the infrared and violet spectra. MFs are most transparent in the red spectrum. In the violet spectrum, light attenuation achieves its maximum. MF transparency increases with proximity to the ultraviolet spectrum.

The magnetic field slightly affects light transmission in properly dispersed MFs in which fine particles that remain are not subjected to aggregation. Large-size particles in a MF promote aggregation owing to magnetic dipole interaction. The latter, as a rule, decreases transparency with the field on; its increase may also be observed, however. In aggregated fluids, dichroism (difference in the attenuation of beams polarized parallel with or normal to the field) and anisotropy of light scattering are very essential.

The magnetic field imparts to MFs the properties of a uniaxial birefringent crystal with its axis aligned with the field vector. Optical anisotropy increases with the field and achieves saturation in the fields of about 200 kA/m. The magneto-optical effect in magnetic colloids is stipulated by the ordering of the orientation of anisotropic ferromagnetic particles and is more pronounced that in molecular liquids.

Acoustical Properties

In practice, rather limited volumes of MFs are usually dealt with. Therefore, the data reported in literature mainly concern the ultrasonic frequency range. In this connection we should mention the two aspects which make MFs attractive to the specialists in acoustics.

Firstly, MFs may serve as an acoustical-magnetical transducer. The variable magnetic field excites a periodic coercive force which deforms the fluid. In case of the field frequency and the natural intrinsic frequency of the fluid match, the acoustic signal is excited by resonance.

Secondly, the field affects the propagation of ultrasound. According to some data, for instance, relative variation of the sound velocity, with the field applied, may be as high as 0.5. On the other hand, in case of well stabilized and monodispersed fluids, the magnetic field does not practically tell on the velocity, and the absorption coefficient increases insignificantly. The variation of acoustical properties in the field is probably characteristic of the MFs subjected to structurization owing to a magnetic diffusion of particles. In a nonuniform field the ferrophase concentration is distributed nonuniformly and so are the acoustical characteristics. Therefore, the aggregation and stratification of particles create the necessary prerequisites for manufacturing acoustic MF-based lenses.

The ferrophase is responsible for specific acoustical effects without magnetic field as well. In concentrated MFs, for example, the sound velocity increases and the absorption coefficient grows (its growth is slower than square-law growth) when the frequency increases. With an increase in temperature, the rate of dispersion decreases, which points to the contribution of interparticle interaction in these phenomena.

1.3 THERMOMECHANICS OF MAGNETIC FLUIDS

In a magnetic field, ferromagnetic particles suspended in a nonmagnetic fluid experience the action by forces and couples which draw them into translational and rotational motion. The ferromagnetic phase interacts with the carrier liquid through viscous friction. Establishing the laws of equilibrium, motion and heat transfer is one of the most important problems in MF studies. In order to solve this problem, it is necessary to reveal the specific features of the momentum exchange, moment-of-momentum exchange, and the energy exchange between the fluid and the field; to study the field effect on transfer processes in MFs; to build mathematical models which correctly allow for these mechanisms; to use these models for the effects of continual fluid-field interaction. Because of complex and diverse physicochemical properties of MFs, these problems are still far from being solved and require consistent efforts on the part of theorists and experimenters.

Quasi-Stationary Approximation of MF Thermomechanics Equations

Basic results of the MF mechanics underlying the majority of applications were derived with quite a simple approach formulated by Neuringer and Rosensweig in 1964 [36]. These investigators suggested that MFs be treated as one-component one-phase fluids whose magnetization is equilibrium in any processes, i.e., it is only determined by the local thermodynamic parameters and field intensities even if these parameters are space- and time-dependent. In this connection, the system of thermomechanic equations for a MF with equilibrium magnetization is also referred to as a quasi-stationary approximation. It is assumed that the field does not affect transfer processes; there is no moment-of-momentum exchange between the fluid and the field; energy is exchanged by the magnetocaloric effect. The main mechanism of interaction-momentum exchange is allowed for by the introduction of volume magnetic force into the equation of motion. The assumption underlying the quasi-stationary approximation holds for well-stabilized MFs based on Newtonian fluids at low and moderate concentrations of magnetically soft particles (magnetic moment may rotate relative to the particle body). This model allows one to calculate the equilibrium shape of the MF surface, the forces pulling a MF into the region of nonuniform fields and expelling nonmagnetic bodies (floating in MF) out of these regions. The most important specific features of dynamic phenomena, such as thermomagnetic convection, propagation of surface and subsurface waves, are considered within the framework of this model.

The quasi-stationary magnetic force induces effects which are in many respects similar to the effects of fluid-gravitational field interaction. Thus, the expulsive force acting upon nonmagnetic bodies is similar to the buoyancy force; thermomagnetic convection is an analog of thermal gravitational connection. The essence of magnetic interaction is that a complex field of volume forces may be created and the geometry of this field may depend on liquid motion and that surface forces may occur. Consistent formulation and some correlations for the quasi-stationary approximation were made in [10, 37]. The quasi-stationary approximation is discussed in the first part of the present book.

Low-Frequency Approximation of MF Thermomechanic Equations

The assumption that there are magnetically soft particles is an extreme idealization in some cases. For example, a rigid dipole model may provide a more adequate notion of the magnetic properties of small cobalt particles. For magnetite particles, the magnetic moment-particle binding energy is also sufficient to effect correlation between rotations of the particle and its magnetic moment. This results in the observation in MFs dynamic phenomena due to the deviation of magnetization from equilibrium, which cannot be interpreted through quasi-stationary approximation [38, 39]. In order to describe these phenomena, the MF thermomechanics equations must take account of the local moment-of-momentum exchange between the fluid and the field. The second part of this book deals with the effects of dynamic interaction within the model suggested in [40]. Just as in the quasi-stationary approximation, the MF is considered as a one-component, one-phase continuum. The moment-of-momentum exchange is taken account of, by introducing volume force couples into the equation of fluid motion. Additional sources of dissipative heat release appear in the energy balance equation. In contrast to other available models, this one does not require any complex MF thermomechanics equations. The application of the constitutive equation which is linear in dynamic parameters, to describe the departure of magnetization from equilibrium, simplifies the model. This ensures a breakthrough in the study of continual effects of the dynamic interaction for nontrival flow and magnetic field geometries.

The relative contribution of the dynamic interaction against the quasi-stationary model is determined by a dimensionless parameter $\omega\tau$ where ω is the characteristic process frequency; τ the magnetization relaxation time of the order of magnitude 10^{-5} s. The linear equation for the dynamic magnetization component can be substantiated microscopically if this parameter is small, i.e., if the characteristic shear rate, macroscopic vorticity and the field frequency are lesser than τ^{-1}, which is valid for various processes. This enables one to characterize the present model as a low-frequency approximation of the MF thermomechanics equations.

The intensity of dynamic interaction, as compared to viscous forces for MFs with magnetically rigid particles, is characterized by a dimensionless parameter φ_h (volume fraction of ferromagnetic particles with a stabilizing shell), whose real estimate may be 0.5. Hence, this mechanism may play an important role even in a low-frequency approximation when the parameter $\omega\tau$ is small. At times its behavior is rather unusual, since it has no analogs either in the dynamics of a normal viscous fluid or in magnetic hydrodynamics. These problems have been omitted in earlier monographs; therefore, we elucidate these problems in this book in greater detail.

The linear hypothesis yields an infinite growth of the dynamic magnetization component with characteristic frequencies. It is clear, however, that such growth has a limit, since the departure of magnetization from equilibrium cannot exceed equilibrium magnetization. Essential deviations from a linear dependence may be expected in case of high-rate shear flows with high-frequency fields when $\omega\tau \lesssim 1$. There are available equations and even systems of equations using the internal degrees of freedom to model the magnetization dynamics in a wide frequency range [14, 41–43]. However, the behavior of nonlinear processes greatly depends on the material of particles and their size. Therefore, the dynamic interaction of real magnetic fluids, with polydispersity being significant and its characteristics often unknown, is more difficult to describe at higher frequencies.

Dipole-dipole interaction may disturb the linearity of the equation and complicate the dynamic properties of MFs at low frequencies. The role of the dipole interaction of particles becomes more important as their size grows [13, 44].

The disturbed concentration homogeneity of the ferromagnetic phase is a significant factor which determines the dynamic interaction, e.g., rotational convection (fluid induced into motion by a revolving field). This factor may be taken into account within the models regarding a magnetic fluid as a multicomponent medium [5, 41].

The study of the specific features in the mechanical behavior of MFs owing to high-frequency demagnetization, aggregation and stratification of particles is an urgent problem to be solved at present. These problems, however, are not the concern of our book which aims at giving the fundamentals of the MF thermomechanics.

1.4 BASIC APPLICATIONS OF MAGNETIC FLUIDS

Magnetofluid Seals

The fact that MFs may be drawn into a strong magnetic field and preserve their reciprocal viscosity predetermines the wide use of MFs in different engineering devices. For example, a magnetofluid seal (MFS) designed to pressurize movable bushings is most widely applied in engineering devices. The MFS design is simple. A highly nonuniform magnetic field is set up in the gap between the casing and a rotating or reciprocating shaft, and retains the MF in the gap. High-quality sealing and low friction moment are the main MFS advantages. A single-stage MFS can maintain a differential pressure up to 0.1 MPa (1 atm). Such characteristics are achieved by using concentrated MFs with 60–70 kA/m saturation magnetization; the magnetic field is focused with the aid of pole tips in a narrow region (2–3 mm wide along the shaft) to result in 1–1.5 Tesla, which is close to saturation of the magnetic circuit material. Multi-stage seals are used to increase the maintained differential pressure. In this case, however, the length of the device is increased to correspond to the number of stages. The MFs design requires a rather high degree of accuracy, because for a strong magnetic field to be established in the gap to be sealed, the latter should be 0.1 mm wide and appropriate requirement should be imposed on the permissible shaft end play. With such a small gap, at higher shaft rotational velocity the fluid in the gap is heated and evaporates, thereby making the seal inoperative. Because of this, the conventional MFS designs may be applied at shaft surface velocities up to 20–30 m/s (depending on the heat transfer condition). The range of permissible velocities may be increased by applying different systems to cool the fluid by refilling in due time and by seeking adequate engineering decisions in order to reduce the relative velocity of the shaft surface and the pole. The use of low-evaporating silicon-based liquids or vacuum oils ensures long service life of magnetofluid seals at moderate velocities of the shaft surface (5–10 m/s) and renders scheduled maintenance unnecessary.

Magnetofluid seals may be used not only for sealing the rotating or reciprocating shafts, but also for protecting engineering devices (bearings, for instance) against contamination. Particles of dust, sand and mud are expelled out of the MFS gap as any nonmagnetic body. The efficiency of such protection is rather high because the magnetic field nonuniformity in the MFS achieves 10^9 A/m^2 and the expulsive force is 10,000 times as large as the particle weight.

Magnetic Fluid as a Lubricant

The use of lubricants increases the operational reliability and service life of machines and mechanisms. However, it is often impossible to retain the lubricant at the contact of moving parts because of its ejection by centrifugal forces or the geometry of the friction unit which does not allow an oil crankcase to be employed. The magnetic field may retain the lubricant at the contact of friction surfaces. As the lubricant efficiency is determined by its viscosity and capability of forming an adsorption layer on friction surfaces, magnetic lubricants are more effective than the ordinary ones even in the absence of a magnetic field. This is because magnetic particles are attracted by steel surfaces thus increasing both the thickness of the absorption layer and the viscosity of the boundary layer. The testing shows that magnetic lubricants, based on silicone fluids and mineral oils, possess high antiscuff properties and may be applied at a temperature up to 300°C, their properties in an external magnetic field being improved.

Removal of Oil Products from Industrial Effluents

The MF properties make it possible to use them for removing oil products from industrial effluents. For instance, when washing tankers of the remains of oil products, the contaminated water must be purified before being drained into the sea, as modern international standards do not allow more than 15 mg of oil products per 1 l of sewage waters. Hydrocarbon-based magnetic fluids, e.g., kerosene-based, dissolve in oil products to transform them into a weak MF. A flow of water containing low-magnetic oil products passes through a magnetic separator which generates a high-gradient magnetic field separating oil products from water. After the separation of the magnetic fluid and oil products, the produced magnetic fluid of high concentration is diluted and again used for purification.

Oil-contaminated surfaces, such as the water areas of harbors and emergency spillings, may be decontaminated in a similar way. An oil-dissoluble magnetic fluid is sprayed on the water surface and then attracted with a magnetic device on board a ship.

Separation of Nonmagnetic Particles with Magnetic Fluid

The additional expulsive force, dependent on fluid magnetization and magnetic field nonuniformity, which exists in a magnetic fluid, has promoted the design of some devices in which this phenomenon is applied. One of them, a magnetohydrostatic separator, is employed to separate the mixtures of nonmagnetic particles according to their density. By choosing the expulsive force strong enough for the particles of a prescribed density to float, one may separate the required fraction from the multicomponent dispersion. To ensure a high degree of separation accuracy, the electromagnet poles are made hyperbolic, which assures a constant field intensity gradient in the volume of magnetic fluid suspended in the interpole space. The separator is installed in an inclined position so that the mixture of particles coming from the bunker move in the magnetic fluid along the poles and are concurrently separated into two fractions. The heavy fraction passes through the layer while the lighter one floats up and slips off the surface into the container. The testing of experimental plants shows that the separators can be used most effectively for 1–6 mm particles. In the separation of non-ferrous metals, such as copper, aluminium, and lead, the accuracy of separation is 0.1–0.2 g/cm^3, the contamination of the components being not more than 0.5%. The capacity of experimental plants amounts to 300 kg/h. Washed from the products of separation, the magnetic fluid is returned to the separator, which reduces the loss amounting to 0.1–0.2 kg per ton of the starting material. Low-concentration magnetic fluid (saturation concentration about $\sim$10 kA/m) is used in the separator; therefore, it is comparatively cheap. The uninterrupted process and nontoxicity of the magnetic fluid (in contrast to a majority of flotation agents) make the magnetohydrostatic separation a promising production process.

Magnetostatic Supports and Bearings

The magnetostatic expulsive force is also used to manufacture axial and radial bearings and supports. In this case, the basic engineering problem to be solved is the establishment of a highly nonuniform field in rather a large volume. Magnetostatic bearings have low friction and noise level; they are to work at high shaft rotational velocities for a long time, exclude the wear-and-tear of friction surfaces and possess damping properties.

Measuring Instruments

The operating principle of a number of measuring instruments is based on the ability of heavy bodies to float up. This is, first of all, a density meter for nonmagnetic bodies with the same principle of operation as that of a separator; by increasing the electromagnet current, a solid particle is made to float up. The required current intensity is the particle density.

The design of an acceleration meter is similar. In this meter a magnetic sphere is retained in the center of the spherical volume of magnetic fluid wherein a magnetic field is set up. The intensity of this magnetic field grows with a distance away from the centre. Such a field is set up by three pairs of electromagnets connected back-to-back and located along three perpendicular axes. As the acceleration grows, the magnetic sphere tends to deviate from the center. The acceleration is determined by the value of this deviation or by the current required to return the sphere. The instrument reading can be 5×10^{-5} g accurate.

A thin magnetic fluid layer can be used as a magneto-optical transducer, since polarization of the transmitted light is determined by the magnetic field. In order to detect internal structural defects, the transducer made as two flat glasses with MF in between, is placed on the surface of the article to be tested. The article is magnetized, and the defect field is observed through crossed polaroids.

The operation of pressure, oscillation and vibration transducers is based on measurements of the force attracting the magnetic fluid to the electromagnet poles.

Magnetic Fluid as a Heat Carrier

As thermomagnetic convection may appreciably exceed natural convection, the application of a magnetic fluid as heat carrier is efficient in the devices which already have strong magnetic fields. Therefore, the replacement of the cooling oil in high-power transformers by a magnetic fluid based on it may considerably increase the permissible transformer loads.

The power of electric motors with prescribed sizes is also determined by the heat removal. When the MF fills the frontal parts of the stator windings and the gap between the stator and the rotor, the thermomagnetic and rotational convections provide additional heat removal.

The power of loudspeakers is limited by the heat removal from the speech coil, since the gap between the coil and the permanent magnet around it is made small to ensure intensive interaction of the magnet and coil fields. The magnetic fluid filling the gap provides intensive heat removal, which allows the permissible load to be increased twice. Meanwhile, the magnetic fluid does additional work; it centers the coil, decreases its nonaxial motion, damps undesirable oscillations thus improving the frequency-response characteristic of the loudspeaker. It is reported that some companies (Telefunken, Philips) produce magnetic fluid loudspeakers.

Magnetic Fluid in Chemical Engineering

The possibility of controlling the shape and stability of the force MF surface makes it a promising technological medium in various mass exchangers, such as spray, packed columns and extractive-distillation towers. The extraction fluid moves as a film or drop flow. In a film extractor employed, for example, to purify associated petroleum gas from heavy impurities, the kerosene film flows down a vertical surface countercurrent to the gas. The capacity of such an apparatus is limited due to separation of the liquid film at high velocities of the gas flow. When using as adsorbent a kerosene-based magnetic fluid flowing down the surface of permanent magnets, the ultimate gas flow velocity is 5 to 7 times as high and achieves 40 m/s. This enables one to diminish the absorber dimensions and to use a forward flow in its vertical-version design, which greatly lowers the cost of the apparatus.

In the course of liquid-liquid extraction, the mass transfer rate is proportional to the total surface of the extractant drops. But the drop deposition velocity in gravity-assisted extractors is in proportion to $\Delta\rho g d^3$ (where $\Delta\rho$ is the difference of liquid phase densities and d—the droplet size). Therefore, although droplets work well for the extraction rate, they can hardly be separated from the dispersing medium, which imposes restrictions on the minimum droplet size. If the extractant is a magnetic fluid, the media can easily be separated by a nonuniform magnetic field. Besides, the size of a droplet in a magnetic field can be increased due to through elongation and internal circulation can be enhanced due through surface vibrations. Both factors promote the mass transfer rate.

Printers

The IBM developments using magnetic fluid as ink have been widely advertised of late. There are several types of printers using magnetic inks. In oscillograph-type devices, the MF jet is deflected by two solenoid pairs, horizontal and vertical. Such a system is able to record analog signals at frequencies up to 600 Hz. If an image needs be reproduced, it is digitized so that logical 0 corresponds to a dark image point, and logical 1 to a light point. Such an image can be reproduced by way of drop printing. The sheet, on which the image is reproduced, is rotating on an axially moving drum 80 that an MF drop may get to any of its points. In the presence of logical signal '1', the magnetic fluid is deflected leaving a light spot on the paper. In the case of logic, signal '0', a drop falls on the paper leaving a black point. The printing density is 15 points per millimeter, which makes a high-quality image.

In another version, drops are extracted from a capillary by an electromagnet field. This ensures an extremely high printout rate: the drop repetition frequency may be 5 kHz.

Flow Control and Drag Reduction by Magnetic Fluid

A magnetic fluid may be an effective means of controlling the flow of ordinary fluids and reducing hydraulic resistances. Thus, in case of a body in a flow, a region of return motion with lower pressure is formed. This results in higher resistance in the flow. There exist techniques of eliminating the separation of the flow from the body surface and the formation of eddies, but all of them are energy-intensive and require the use of additional equipment. Coating the streamlined body with a layer of low-viscosity magnetic fluid retained by a magnetic field reduces shear stresses at the flow boundary. The flow entrains the surface layers of the magnetic fluid, causing it to circulate in the layer, and, in a manner, slides on them. The region of return motion is therefore greatly decreased and, hence, the resistance of the body moving in the flow is reduced. When high-viscosity fluids are piped, the resistance may also be reduced by creating a wall boundary layer of a low-viscosity magnetic fluid. The discontinuity of the layer prevents its wear. The viscosity ratio of magnetic and nonmagnetic fluids being 1:100, a tenfold gain in resistance may be achieved.

Magnetic Fluids in Medicine

The first commercial magnetic fluids immediately attracted the attention of physicians as a material to be used as artificial clots that might occlude large blood vessels during operations. Absence of mechanical damage due to clamps accelerates the healing of blood vessels in the post-operation period. A certain interest is shown to magnetic fluids as carriers of medicinal preparations. With the aid of a magnetic field, a drug may be transported to the required area, for example, to a tumor, and can be retained there for as long as necessary. Such a method appreciably increases the efficiency of drugs, allows localized cure of the disease focus and protects the organism against side effects. The use of a magnetic fluid as an radiopaque substance is however most developed. The magnetite particles contained in the magnetic fluid are highly opaque to X-rays. Therefore, a magnetic fluid with an 8 to 10 per cent volume concentration of the magnetite is not inferior in absorbing X-rays to such a widely used preparation as barium sulfate. The possibility of introducing radiopaque preparations into hollow organs (gullet, stomach, urinary, bladder etc.), of retaining them in a contracting organ, for example in the gullet, and transporting them along the organ walls irrespective of the gravity direction, markedly expands the X-ray diagnostics application and allows the elasticity of hollow organ walls to be estimated and, thus, the disease to be detected in its early stage.

Application Prospects

A magnetic fluid is a multi-functional medium, making possible its most unexpected applications even irrespective of its magnetic properties. For instance, owing to surfactants and numerous particles carrying electric charges, it is excellent foam-forming additive. Japanese investigators have found that an addition of kerosene- or water-based magnetic fluid to internal-combustion engine fuel in the volume ratio 1:350 reduces the nitrogen oxide content in exhaust gases, by 25 per cent, prevents corrosion of the combustion chamber and increases the combustion efficiency.

It is worth noting in conclusion that the diversity of magnetic fluid applications is far from being exhausted. Investigations of magnetic fluids now under way all over the world promote the expansion of MF applications. This makes magnetic fluids, an artificial medium, ever more attractive to engineers, designers and scientists.

PART ONE

EQUILIBRIUM MAGNETIZATION

CHAPTER

TWO

THERMOMECHANIC EQUATIONS FOR MAGNETIC FLUIDS OF EQUILIBRIUM MAGNETIZATION

INSTANTANEOUS MAGNETIZATION RELAXATION

The main physical prerequisite for the existence of equilibrium magnetization is the assumption that nothing, except thermal motion, hinders the orientation of elementary magnetic moments along the field and that the mean value of magnetization is achieved instantaneously, i.e., within the times much shorter than the characteristic times of macroscopic processes (hydrodynamic, thermal, electromagnetic, etc.). This assumption makes it possible to consider the fluid magnetization vector $\vec{M}$ at a given instant to be parallel to the vector of magnetic field intensity $\vec{H}$, which in the general form may be related as $\vec{M} = (M/H)\vec{H}$. Magnetization M is determined by the fluid temperature and density and by field intensity: $M = M(T,\rho,H)$. It is natural that it decreases with rising temperature and increases with the field intensity.

The condition for the vectors $\vec{M}$ and $\vec{H}$ to be parallel is realized in a MF only for certain colloid characteristics that will be estimated in Ch. Six. Nevertheless, for a wide range of problems this condition may be regarded as fulfilled and enables one to study those effects in a MF which are caused to occur by the volume magnetic force due to the interaction between equilibrium magnetization and the magnetic field.

2.1 MAGNETIC FLUID AS HOMOGENEOUS ISOTROPIC CONTINUUM

Recall that a magnetic fluid consists of small (about 10 nm in dia.) solid ferromagnetic particles dispersed in a carrier liquid. The surfactant shell and the

Brownian motion prevent their coalescence and deposition even in highly non-uniform magnetic fields. As a result of chemical interaction, a layer of substance which does not possess magnetic properties may form on the surface of a solid ferromagnetic particle. For instance, a layer of ferrous oleate is formed on magnetite particles in the solvate shell of olic acid. The layer is about 0.5 nm thick. Thus, it is not the entire volume of a solid particle in the colloid that is ferromagnetic. This requires that two particle radii, namely total radius R_s and ferromagnetic portion radius R_f, be introduced.

Below, if need be, all the values related to the ferromagnetic particle material will have a subscript f and those to the carrier liquid, will have a subscript 0.

The homogeneous continuum approximation implies, first of all, that considered are the processes whose characteristic dimensions are much greater than the distance between ferromagnetic particles. The total magnetic moment of particles is assumed to be equally distributed throughout any elementary fluid volume.

Neglect of Dipole-Dipole Interaction

As the size of particles in a ferromagnetic colloid increases, the magnetic moment of each particle grows and dipole-dipole interaction between them is enhanced. The Brownian motion cannot prevent their coalescence or the formation of chains and clusters. The structurization of the magnetic phase becomes essential and results in non-Newtonian fluid properties typical of ferromagnetic suspensions with particle sizes about 1 μm [44] and in hysteresis of the magnetization curve.

The energy of dipole-dipole interaction E_d between the particles with magnetic moments m being r apart is of the order $E_d \sim \mu_0 m^2/r^3$. In order to have the predominance of the thermal energy $kT > E_d$ for the magnetite particles ($m \sim 10^{-19}$ J/Tesla) of radius $R_s = 4$ nm at room temperatures $kT \sim 10^{-21}$ J, the distance between them should not be less than 20 nm $= 5R_s$ which is achieved by introducing surfactants into the colloid.

The aspects connected with the behavior of non-Newtonian magnetic fluids may be the subject of an independent study, as it is, for example, the case of magnetorheological effect [44]. These aspects are not considered in this book.

Concentrational Homogeneity of Colloid

As for ferromagnetic colloids, one more fact is worthy of consideration. In external fields (gravitational and magnetic), particle concentration is redistributed in a primarily homogeneous colloid. This distribution is related by $\varphi = \varphi_* \exp(-U/kT)$, where U is the potential energy of a particle in the external field. For magnetic particles with a magnetic moment m

$$U = \mu_0 m |\nabla H| l$$

where l is the characteristic size of the system. In such a situation, the colloid magnetization, determined by the volume concentration of ferromagnetic particles, will depend both on the magnetic field intensity and the field gradient. Taking account of the concentration effects requires that the diffusion processes be discussed. Although such attempts were made elsewhere [9], a majority of the results on MF thermomechanics have been obtained by neglecting these effects. The validity of such an approach is ensured by the following. Firstly, a change in the concentration of particles in the magnetic field may be ignored if $U \ll kT$. Hence, for the gradients of the magnetic field wherein the colloid may be regarded homogeneous obtain the condition $|\nabla H| \ll kT/(\mu_0 ml)$. At $kT \sim 10^{-21}$ J, $m \sim 10^{-19}$ J/Tesla and $l \sim 10^{-2}$ m follows $|\nabla H| \ll 10^6$ A/m^2.

Secondly, the equilibrium distribution of the particle concentration proceeds for a finite time τ_c. Its order is specified by the characteristic diffusion time $\tau_c \sim l^2/D$. The diffusion coefficient for spherical particles of a radius R, suspended in a viscous fluid with the viscosity coefficient η is $D = kT/(6\pi\eta R)$ [2]. The characteristic dimension of concentrational nonhomogeneity, l, is determined by the exponent in the Boltzmann distribution $U/kT = fl/kT \sim 1$, where f is the force acting on a particle. In this case, $f = \mu_0 m/\nabla H|$ and $l \sim kT/f = kT/(\mu_0 m|\nabla H|)$. Using these relations, get

$$\tau_c \sim 6\pi\eta RkT/f^2 = 6\pi\eta RkT/(\mu_0 m|\nabla H|)^2$$

At $\eta \sim 10^{-3}$ kg/(m · s); $R_s \sim 5 \cdot 10^{-9}$ m; $kT \sim 10^{-21}$ J; m $\sim 10^{-19}$ J/Tesla; $|\nabla H| \sim 10^6$ A/m^2 follows $\tau_c \sim 6 \cdot 10^6$ s $\sim$ 60 days. If the times of the processes under consideration are much less than τ_c, the concentrational nonhomogeneity of the colloid may be neglected.

Determination of Some Physical Properties of Ferrocolloid by the Properties of Starting Materials

The available handbooks contain rather comprehensive information on the physicomechanical properties of the starting substances which constitute ferrocolloid [45]. Relations will be given below to estimate the macroscopic colloid characteristics using these data. The ferrocolloid is regarded as a set of ferromagnetic particles uniformly distributed in a carrier liquid. Studied are only the colloid properties which are not appreciably affected by surfactants due to their low concentration.

The reference values are: volume concentration φ of ferromagnetic particles in the colloid; particle radii R_s, and R_f; saturation magnetization of ferromagnetic M_{sf}; densities of carrier liquid and ferromagnetic ρ_0 and ρ_f as well as all the other known properties of the colloid constituents. The volume concentration of particles φ is determined by the ratio of their volume to the entire colloid volume.

In a first approximation, the ferromagnetic particle is considered to be spherical. Its total volume V_s is therefore related as $V_s = (4/3)\,\pi R_s^3$, and the volume of the ferromagnetic portion as $V_f = 4\pi R_f^3/3$. The quantity of particles per unit colloid volume $n = \varphi/V_s$.

Assuming the magnetization of an individual particle to be equal to the saturation magnetization of the starting ferromagnetic material, for the magnetic moment of the particle obtain the expression $m = V_f M_{sf}$. Then the colloid saturation magnetization

$$M_s = nm = \varphi(R_f/R_s)^3 M_{sf}$$

The mass of the substance in the colloid volume is the sum of the mass of particles and the mass of carrier liquid. Hence, the colloid density

$$\rho = \rho_f\varphi + (1 - \varphi)\rho_0 \tag{2.1}$$

A similar relation is obtained for the thermal expansion coefficient of the colloid

$$\beta_\rho = -(1/\rho)(d\rho/dT) = \varphi\beta_{\rho f} + (1 - \varphi)\beta_{\rho 0} \tag{2.2}$$

The expression of the same kind holds for the volume heat of the colloid

$$c = c_f\varphi + (1 - \varphi)c_0 \tag{2.3}$$

The colloid viscosity coefficient may be estimated by the Wand formula

$$\eta = \eta_0 \exp\left[\frac{2.5\varphi - 2.7\varphi^2}{1 - 0.609\varphi}\right]$$

which for low concentrations ($\varphi \ll 1$) becomes the known Einstein equation

$$\eta = \eta_0(1 + 2.5\varphi)$$

The thermal conductivity coefficient at small volume concentrations of particles may be estimated from the known relation (cf. [7])

$$\lambda = \lambda_0 + 3\lambda_0\varphi(\lambda_f - \lambda_0)/(\lambda_f + 2\lambda_0)$$

Find the relationship between the quantity of particles and the colloid density. With neglect of the possible generation and decay of particles, we conclude that in case of equilibrium for any containment volume, V, of the colloid, the quantity of magnetic particles, N_f and N_0 nonmagnetic fluid molecules it contains is a constant. Their ratio is also constant and independent of the volume considered

$$N_0/N_f = (N_0/V)(V/N_f) = n_0/n_f = C_{f0}$$

The value of C_{f0} is constant for the specified colloid and determines the number of carrier liquid molecules per one ferromagnetic particle. If the particle mass is w_f, and the mass of one carrier liquid molecule is w_0, then the colloid density is determined as

$$\rho = n_f w_f + n_0 w_0 = n_f(w_f + C_{f0} w_0) = n_f \overline{w}$$

where $\overline{w} = w_f + C_{f0}w_0 = \text{const}$ is the mass of a particle with 'attached' carrier liquid molecules. The expression linearly relates, without a free term the colloid density ρ and the quantity of solid particles, n, per unit volume.

It may seem, at first sight, that with decreasing particle concentration the colloid density must tend to the carrier liquid density rather than to zero, as it follows from the above relation. Such an opinion would be quite realistic should the particle concentration change as a result of the introduction or removal of particles from the colloid due to their generation or decay, which is hardly valid, if at all for a certain colloid. For any particular colloid, i.e., a certain set of solid ferromagnetic particles and liquid molecules being in a state of thermodynamic equilibrium, the particle concentration may only vary with its volume. The concentration of particles as well as the colloid density vanish simultaneously at its infinite expansion, which is represented by the relation $\rho = n\overline{w}$.

A more intricate situation is observed in case of irreversible processes, for example, particle diffusion in the colloid. The discussion of this problem is beyond the scope of our work.

As the saturation magnetization, M_s, of the colloid is also in proportion to the particle concentration, $M_s = nm$, we obtain a linear dependence of saturation magnetization on the density:

$$M_s = (m/\overline{w})\,\rho$$

i.e., $\partial M_s/\partial\rho = M_s/\rho$. This very relation, as will be seen below, also holds for any colloid magnetization.

As an example, let us calculate the properties of the colloidal solution of magnetite particles in kerosene using the above formulae (cf. Table 2.1). Assume the particle radii $R_s = 4 \cdot 10^{-9}$ m, $R_f = 3.5 \cdot 10^{-9}$ m; their volume concentration $\varphi = 10\%$. Then $V_s = 2.7 \times 10^{-25}$ m^3; $V_f = 1.8 \times 10^{-25}$ m^3; $n = 3.7 \times 10^{23}$ m^{-3}; $V_f/V_s = 0.67$.

More particular experimental data on the physical and mechanical properties of some ferrocolloids are give in Table 2.2, taken from [12].

The comparison of the calculated characteristics for 10% ferrocolloid magnetite in kerosene presented in Table 2.1 with the data of Table 2.2 for a hydrocarbon-based ferrocolloid with 31.8 kA/m saturation magnetization (third line at the top) gives their fine agreement within less than 1 per cent error for density

TABLE 2.1

Substance	Saturation magnetization M_s, кA/m	Density $\rho \cdot 10^{-3}$, kg/m^3	Heat capacity $c \times 10^{-6}$, J/(m$^3 \cdot$ K)	Thermal expansion coefficient $\beta_\rho \times 10^4$, K^{-1}
Magnetite (F_3O_4, FeO $\times$ Fe_2O_3)	480	4.8–5.3	3.1	0,27
Kerosene	–	0.82	1.64	10.0
Ferrocolloid	32	1,24	1.79	9.0

and saturation magnetization, less than 3 per cent error for heat capacity and about 4.5 per cent error for the thermal expansion coefficient. This may be considered a good result, and all the more so because the exact characteristics of the starting materials are not reported in [12] and may differ, to some extent, from the values we have taken for calculations.

The magnetic moment of a particle here $m = 0.86 \cdot 10^{-19}$ J/Tesla.

Magnetic Fluid Thermomechanics as a Section in Magnetic Hydrodynamics

Thermomechanics of magnetic fluids with equilibrium magnetization deals with a one-component fluid which is capable of being magnetized by an external magnetic field applied with the magnetization law $\vec{M} = \chi(\rho,T,H)\vec{H}$.

With such a formulation, it is an integral part of such a field of liquid and gas mechanics as magnetic hydrodynamics. Before the advent of MFs, the classical magnetic hydrodynamics only concerned the motion of conductive liquids and gases in electromagnetic fields [3]. It has scored great successes in explaining many natural phenomena, in particular, astrophysical ones and in engineering applications. This refers, above all, to the implementation of thermonuclear and MHD energy generators. The ponderomotive interaction of a conductive fluid and magnetic field is ensured by the conduction currents in it. The interaction may be referred to as magnetodynamic in contrast to magnetostatic associated with continuum magnetization. In continuum electrodynamics, both kinds of interaction are equitable and are described by the Maxwellian stress tensor of an electromagnetic field in the continuum. If should also be noted that while the field strength plays a decisive role for the interaction of conduction currents and the field, then the magnetostatic force associated with fluid magnetization is determined by the field gradient magnetic fluid is drawn into the region of higher field strength). The consideration of both forces ensures a most comprehensive study of the effect of a magnetic field, its constant component and gradients on the mechanical processes in a fluid. In this respect, ferrohydrodynamics makes up for a gap whose elimination would allow magnetic hydrody-

TABLE 2.2

Carrier liquid	Saturation magnetiz. M_s, kA/m	Density, $\rho \times 10^{-3}$, kg/m^3	Viscosity, η, kg/(m · s)	Pour point T, K	Surface tension coeff., α, N/m	Thermal conduct. λ, W/(m · K)	Heat capacity per unit volume $C \times 10^{-6}$ J/(m^3 · K)	Thermal expansion coefficient $\beta_\rho \times 10^{-4}$, K^{-1}
Diesters	15.9	1.185	0.075	236	–	–	–	–
Hydrocarbons	15.9	1.05	0.003	273	0,028	0.15	1.715	9.0
	31.8	1.25	0.006	281	0.028	0.15	1.840	8.6
Fluoro-organic matter	7.96	2,05	2.5	239	0.018	0.2	1.966	10.6
Esters	15.9	1.15	0.014	217	0.026	0.31	3,724	8.1
	31.8	1.30	0.030	217	0.026	0,31	3.724	8.1
	47.7	1.40	0.035	217	0.021	0.31	3.724	8.1
Water	15.9	1.18	0.007	273	0.026	1.4	4.184	5.2
	31.8	1.38	0.01	273	0.026	1,4	4.184	5.0
Polyphenyl esters	7.96	2.05	7.5	283	–	–	–	–

* Solidification point corresponds to viscosity of 100 kg/(m · s)

namics to give a complete picture of mechanical behaviour of fluids in magnetic fields.

Despite a relatively short period of its existence, ferrohydrodynamics has made certain progress in practical implementation of its achievements in a wide range of industries. In order to promote further advances in MF applications, their thermomechanics must be intensively studied so as to give a fresh impetus to the development of magnetic hydrodynamics as a whole.

2.2 EQUILIBRIUM MAGNETIZATION EQUATIONS

Bearing in mind that the approximation under study assumes that the vectors of magnetization $\vec{M}$ and magnetic field strength in the continuum $\vec{H}$ are parallel, the law of magnetization must relate the modules of these vectors.

Langevin Behavior of the Magnetization Curve

In ferromagnetic colloids, ferromagnetic particles are the elementary carriers of a magnetic moment. Their magnetization is of ferromagnetic nature, i.e., it owes to the exchange interaction between the atoms of a substance. Small dimensions of particles ($\sim 10^{-8}$ m) ensure their single-domain character. At the same time, their dilution in the colloid at moderate concentrations weakens their magnetic interaction to an extent (promoted by surfactants) as well that they may be considered as noninteracting Brownian particles participating in chaotic thermal motion with energy kT. As a result, it is possible to represent a set of such particles as a rarefied gas and to describe its magnetization using the theory of paramagnetic gas magnetization [4]. This theory yields the law of magnetization described by the Langevin function

$$M = nm\,(\coth\xi - 1/\xi),\ \xi = \mu_0 mH/kT \tag{2.4}$$

The high values of magnetic moments of particles m give grounds to refer to such systems as superparamagnetics. Their magnetization is in direct proportion to the quantity of magnetic particles per unit continuum volume n. The Langevin function is plotted in Fig. 2.1. With increasing magnetic field intensity ($\xi \to \infty$), the curve asymptotically approaches a unity, which consists with the continuum saturation magnetization $M_s = nm$, i.e., with complete orientation of all the particles along the field.

Using the linear relationship between the concentration of colloid particles and its density $\rho = \overline{w}n$, magnetization law (2.4) may be written in the form

$$M = \rho(m/\overline{w})(\coth\xi - 1/\xi) \tag{2.5}$$

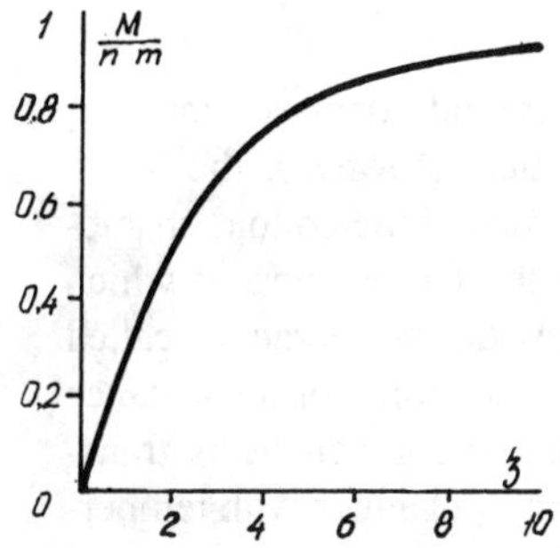

FIG. 2.1

to indicate a linear dependence of the colloid magnetization on its density ρ at $\partial M|\partial\rho = M\rho$.

In accordance with the Langevin curve, saturation magnetization is practically achieved at $\xi \sim 10$. At room temperatures and $m \sim 10^{-19}$ J/Tesla, this corresponds to magnetic field intensity $H \sim 10^5$ A/m. For comparison, Fig. 2.2 gives a magnetization curve for ferromagnetic fluid MK-26 being a colloidal magnetite solution in kerosene with volume concentration of a solid phase ~8 per cent.

The initial segment of magnetization curve (2.4) corresponding to low values of the argument ξ, i.e., to low magnetic field intensity $H \ll kT/(\mu_0 m)$, is linear. At $\xi \ll 1$ (2.4) gives

$$M = nm\xi/3 = \mu_0 nm^2 H/3kT \qquad (2.6)$$

and for magnetic susceptibility stems the Curie law

$$\chi = M/H = \mu_0 nm^2/3kT = C_A/T \qquad (2.7)$$

The quantity of particles per unit volume of the ferromagnetic colloid is of the order 10^{17} cm^{-3} = 10^{23} m^{-3}. At $m \sim 10^{-19}$ J/Tesla and room temperatures, we obtain from (2.4) and (2.7) $M_s \sim 10^4$ A/m and $\chi \sim 1$, which corresponds to real fluids.

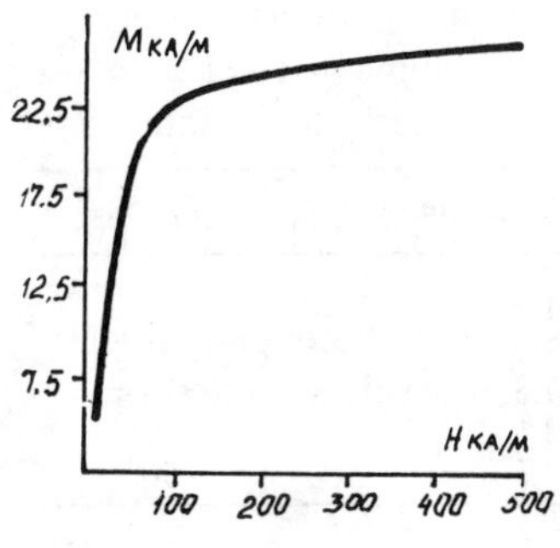

FIG. 2.2

The Curie Point

The Langevin-type magnetization of ferromagnetic colloids deprives them of hysteresis characteristic inherent in solid ferromagnetics. However, the ferromagnetic nature of solid particles exposes some properties of the colloid characteristic for ferromagnetics. First of all, they include the temperature at which strong magnetic properties of a substance disappear. At this temperature, called the Curie point T_K, the exchange interaction between the atoms of a substance disappears, the domain structure is destroyed and the ferromagnetic is transformed into a paramagnet whose magnetic susceptibility χ changes with temperature by the Curie-Weiss law

$$\chi = C_K/(T - T_K) \tag{2.8}$$

Tables 2.3 and 2.4 present the values of saturation magnetization M_{sf} at 20°C, specific (per unit mass) magnetization at 0 K $-$ σ_0 and Curie point T_K for some ferromagnetics [4, 45].

It is seen from these Tables that the Curie point for ferromagnetics ranges widely from negative to some hundreds of Centigrade degrees above zero, depending on the substance. Of special interest are the substances for which the Curie point is close to room temperature. They include, for instance, gadolinium and some alloys represented in Table 2.4 by 30% permalloy.

The disappearance of ferromagnetic properties of colloid particles at the Curie point leads to the disappearance of its magnetization, i.e., it becomes an ordinary paramagnetic material.

Temperature Dependence of Ferrocolloid Magnetization

The ferromagnetic properties of a substance in colloid are determined by the particle magnetic moment m which in a first approximation is related to the solid ferromagnetic magnetization M_{sf} as $m = M_{sf}V_f$, where V_f is the volume of the ferromagnetic portion of a particle. Near the Curie point, the spontaneous ferromagnetic magnetization decreases in proportion to $\sqrt{T_K - T}$ [3]. The particle moment, too, obeys the same law and tends to zero $m = \alpha_K\sqrt{T_K - T}$. As formulae (2.6) and (2.7) are also valid, the colloid magnetization and its magnetic

TABLE 2.3

Substance	M_s(20° C), KA/m	σ_0 (0 K), A · m³/(m · kg)	T_K, °C
Fe	1714	221.9	770
Co	1422	162.5	1331
Ni	484.1	57.5	358
Gd	0	253.5	20
Tb	0	173.5	−50

TABLE 2.4 Curie Point for Some Alloys

Alloy	78% Permalloy (22% Fe and 78% Ni)	30% Permalloy	Magnetic (31% FeO, 69% Fe_2O_3)
T_K °C	550	70	550–600

susceptibility near the Curie point tend to zero in proportion to the difference $(T_K - T)$:

$$\left.\begin{aligned} M &= (\mu_0 n a_K^2/3kT)(T_K - T)H = (\mu_0 n a_K^2/3k)(T_K/T - 1)H; \\ \chi &= (\mu_0 n a_K^2/3k)(T_K/T - 1) \end{aligned}\right\} \tag{2.9}$$

Fig. 2.3 presents the saturation magnetization vs. temperature curves for some ferromagnets [4]. Although there is no exact formula that might describe this curve for the entire temperature range, it is nevertheless approximated rather well by the following transcendental equation [4]:

$$M_{sf}/M_{sf}^0 = \text{th}\,\frac{M_{sf}/M_{sf}^0}{T/T_K} \tag{2.10}$$

where M_{sf}^0 is the saturation magnetization of a substance at 0 Curie.

Assuming that the magnetic moment of a separate single-domain particle is proportional to the saturation magnetization of solid ferromagnet $m = M_{sf}V_f$, we arrive at

$$m/m_0 = \text{th}\,\frac{m/m_0}{T/T_K} \tag{2.11}$$

where $m_0 = M_{sf}^0 V_f$ is the particle moment at $T = 0$ Curie. This equation is also represented by curve 1 in Fig. 2.3.

In ref. [4], a more convenient approximation relation is given for calculating the dependence of saturation magnetization of a ferromagnet on temperature near the Curie point

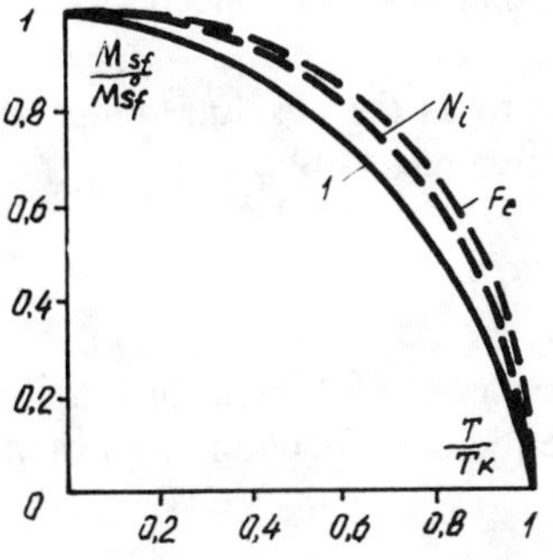

FIG. 2.3

$$M_{sf} = M_{sf}^0[3(T_\kappa - T)/T_\kappa]^{1/2} \tag{2.12}$$

For the magnetic moment of a separate particle we shall write

$$m/m_0 = [3(T_\kappa - T)/T_\kappa]^{1/2} \tag{2.13}$$

Temperature dependence of the magnetic moment of particles is most distinct in the vicinity of the Curie point. Far from the Curie point, the temperature dependence of the colloid magnetization is mainly determined by two other factors.

One is the thermal disorienting motion of particles which is accelerated as the temperature grows. It is explicitly described by the temperature in the argument of Langevin function (2.4): $\xi = \mu_0 mH/kT$.

The other is associated with the fact that higher temperature of the fluid causes its volume to increase and, hence, its magnetization to decrease.

The temperature dependence of the particle concentration is specified by the temperature dependence of the colloid density

$$n(T) = (1/\overline{w})\rho(T) \tag{2.14}$$

Equations (2.14), (2.10)–(2.13) together with (2.4) and (2.5) give comprehensive information about the dependence of ferromagnetic colloid magnetization on the temperature in its wide range of variations.

Linear Equation of Magnetized State

For small temperature and magnetic field intensity ranges, the equation of magnetized state for incompressible fluid $M = M[\mathrm{T,H},\rho(T)]$ may be approximated by means of linear relations by expanding this function in terms of the degrees of temperature and field intensity deviations from certain equilibrium quantities T^* and H^*:

$$M = M^* + (dM/dT)_{\mathrm{H}}(T - T^*) + (dM/dH)_T(H - H^*) \tag{2.15}$$

Here $M^* = M(T^*,H^*)$; $(dM/dH)_T = \chi_r$ is the differential magnetic susceptibility of the fluid. Along with this quantity, we shall use the concept of total or integrated magnetic susceptibility $\chi_s = M/H$.

The quantity $K = -(dM/dT)_H$ is referred to as the absolute temperature magnetization coefficient in contrast to the relative one

$$\beta_M = -(1/M)(dM/dT)_H$$

As estimated above, the temperature dependence of MF magnetization is determined explicitly; by the particle magnetic moment function $m(T)$, and by the fluid thermal expansion $\rho(T)$. Therefore,

$$(\partial M/dT)\rho = (\partial M/\partial T)_{H,\rho,m} + (\partial M/\partial m)_{H,\rho,T}(\partial m/\partial T)_\rho + (\partial M/\partial \rho)_{H,T,m}(\partial \rho/\partial T) \quad (2.16)$$

The last term in this relation is most easy to calculate. With regard for (2.5)

$$(\partial M/\partial \rho)_{H,m,T}(\partial \rho/\partial T) = M(1/\rho)(\partial \rho/\partial T) = -M\beta_\rho$$

where β_ρ is the fluid thermal expansion coefficient determined by (2.2).

The first two terms in (2.6) are easily evaluated in two limited cases: 1) in a state of saturation ($\xi \gg 1$); 2) in the initial linear section of the magnetization curve ($\xi \ll 1$).

In each of these cases, we have from (2.5).

1) $\xi \gg 1$; $M = \rho(m/\overline{w})$; $(\partial M/\partial m)_{H,\rho,T}(\partial m/\partial T) = M(1/m)(\partial m/\partial T) = -M\beta_{Mf}$ where $\beta_{Mf} = -(1/m)(\partial m/\partial T)$ – is the relative temperature coefficient of the magnetic moment of a separate particle.

In the instant under consideration, with the fluid in a state of saturation and all its particles oriented along the field despite the thermal Brownian motion, it does not naturally contribute to thermal demagnetization of the fluid $(\partial M/\partial T)_{H,m,\rho} = 0$.

In this case,

$$K = M(\beta_{Mf} + \beta_\rho);\ \beta_M = \beta_{Mf} + \beta_\rho \quad (2.17)$$

Far from the Curie Point (for magnetites at room temperatures), the temperature dependence of the magnetic moment of particles is very weak $\beta_{Mf} \approx 0$. In a state of fluid saturation and far from the Curie point, this means that the temperature dependence of magnetization is only determined by the thermal expansion $\beta_M = \beta_\rho$ and is of the order $(10^3-10^{-4})\ K^{-1}$. At $M \sim (10^4-10^5)$ A/m, we obtain $K \sim (1-10^2) \cdot A/(m \cdot K)$.

Of peculiar interest is the fact associated with unique properties of water at (0–4)°C. In this temperature range, its thermal expansion coefficient is negative $\beta_{\rho 0} < 0$, which may provide a negative value of the thermal expansion coefficient of a water-based MF ($\beta_\rho < 0$) in accordance with (2.2). This, in turn, implies negative temperature coefficient of fluid magnetization $\beta_M < 0$, i.e., with increasing temperature, MF magnetization may grow, owing to compression of the carrier liquid.

The numerical values of the temperature magnetization coefficient of particles β_{Mf} are found from the temperature dependence for their material

$$\beta_{Mf} = (1/m)(\partial m/\partial T)_\rho = (1/M_{sf})(\partial M_{sf}/\partial T)_\rho$$

by formulae (2.10) to (2.13) or by experimental relations in Fig. 2.3. In the vicinity of the Curie point its values may achieve $\beta_{Mf} \sim 10^{-2}\ K^{-1}$ and more, and its contribution to the temperature magnetization coefficient of the fluid will be the decisive one.

2) In the other limiting case (the initial section of the magnetization curve) $\xi \ll 1$ it follows from (2.5)

$$M = (\mu_0 \rho / 3k\overline{w}) \cdot (m^2/T)H,$$

hence

$$(\partial M/\partial T)_{H,\rho,m} = -M/T; \; (\partial M/\partial m)_{H,\rho,T}(\partial m/\partial T) = -2M\beta_{Mf}$$

Finally:

$$K = M(1/T + 2\beta_{Mf} + \beta_\rho); \; \beta_M = 1/T + 2\beta_{Mf} + \beta_\rho \qquad (2.18)$$

As compared to the above case, the temperature magnetization coefficient includes a new term $1/T$ for thermal disorienting motion of the magnetic moment. At room temperatures, $T = 300$ K, this term, $3 \times 10^{-3}\ \mathrm{K}^{-1}$, is determining ($\beta_M \approx 1/T$). In other respects, all the considerations for the above case of fluid in a state of saturation are valid here.

2.3 THERMODYNAMIC RELATIONS. MAGNETOCALORIC EFFECT

The basic relation in the thermodynamics of magnetics is the one that specifies the work δW done on the magnetic substance with an infinitely small change of the magnetic field $\delta \vec{B}$ in it. This work is responsible for the changes of the internal energy of the body, including the field energy, at constant entropy S and is determined by the relation $\delta W = \int \vec{H} \delta \vec{B} dV$ [3]. The work based on a unit volume is written as: $W = \vec{H} d\vec{B}$. With this in view, the differentials of internal energy dU and dF free energy $= d/U - TS)$ based on a unit volume, are of the form

$$dU = dU_0(\rho,S) + \vec{H} d\vec{B}; \; dF = dF_0(\rho,T) + \vec{H} d\vec{B} \qquad (2.19)$$

Here U_0, F_0 are the internal and free energies of the unit volume of the system, with its density and temperature at play and in the absence of a magnetic field: $dU_0 = TdS + \xi d\rho$, $dF_0 = -SdT + \xi d\rho$ (where ξ is the chemical potential of the continuum unit volume).

Of great significance are the thermodynamic potentials $\tilde{U} = U - \vec{H}\vec{B}$, $\tilde{F} = F - \vec{H}\vec{B}$, the differentials of which are determined by the expressions:

$$d\tilde{U} = dU_0 - \vec{B} d\vec{H}; \; d\tilde{F} = dF_0 - \vec{B} d\vec{H} \qquad (2.20)$$

While the former (dU, dF) stand for the work performed at constant magnetic field potentials, the latter ($d\tilde{U}$, $d\tilde{F}$), at constant sources (currents). At equilibrium magnetization, $\vec{H} \parallel \vec{B}$ and $\vec{H}d\vec{B} = HdB$

Stress Tensor in Magnetic Fluid

When we know the free energy $\tilde{F}$, we are able to write the expression for the stress tensor in a magnetic fluid [3]

$$\sigma_{ik} = [\tilde{F} - \rho(\partial\tilde{F}/\partial\rho)_{T,H}]\delta_{ik} + H_i B_k \tag{2.21}$$

With regard to (2.20), we have

$$\sigma_{ik} = -P(\rho,T)\,\delta_{ik} - \mu_0\delta_{ik}\int_0^H [M - \rho(\partial M/\partial\rho)_{T,H}]dH - \mu_0(H^2/2)\delta_{ik} + H_i B_k \tag{2.22}$$

Here $P(\rho,T) = F_0 - \rho(\partial F_0/\partial\rho)_T$ – is the fluid pressure determined by the equation of state which holds in the absence of a magnetic field at appropriate values of density and temperature.

For fluids with Langevin magnetization, M is proportional to the quantity of magnetic particles per unit volume or to density ρ cf. (2.5). Therefore, $\rho(\partial M/\partial\rho)_{T,H} = M$, and the second term in (2.22) vanishes identically.

Taking account of the viscous stresses specified by the tensor σ'_{ik}, the stress tensor in a magnetic fluid acquires the final form

$$\sigma_{ik} = -P\delta_{ik} + \sigma'_{ik} - \mu_0(H^2/2)\delta_{ik} + H_i B_k \tag{2.23}$$

and, in addition to the stresses occurring in an ordinary liquid, also includes the Maxwellian stress tensor for a magnetic field in continuum.

Magnetocaloric Effect

The work done to perform adiabatic magnetization and demagnetization of a magnet causes the temperature of the latter to change [4]. The adiabatic application of the magnetic field involves magnetization of the magnetic with the result that in its temperature rises. With the magnetic field off, the magnetic is demagnetized and its temperature drops. This phenomenon is referred to as a magnetocaloric effect.

This phenomenon may be considered within the thermodynamics using thermodynamic equalities (2.20). For instance, it follows from the fact the change of the internal magnetic energy $d\tilde{U} = TdS + \xi d\rho - BdH$ is a total differential

$$(dT/dH)_{S,\rho} = -(dB/dS)_{\rho,H} = -\mu_0(dM/dS)_{\rho,H} \tag{2.24}$$

In the instant under consideration $dS = dQ/T = c_{\rho,H}dT/T$, where $c_{\rho,H}$ is the heat capacity of the magnetic unit volume; dQ is the quantity of heat generated in unit volume. With this in view, (2.24) gives a relation for the magnetic temperature change with dT the field in it

$$dT = -\frac{\mu_0 T}{c_{\rho,H}}\left(\frac{dM}{dT}\right)_{\rho,H} dH \tag{2.25}$$

Magnetization of the magnetic material decreases as the temperature increases; therefore, $(dM/dT)_{\rho,H} < 0$. For ferromagnetics $|dM|dT|$ has its highest value near the Curie point, the magnetocaloric effect being maximum at this point. Fig. 2.4 gives experimental curves for the magnetocaloric effect in iron, with magnetic fields of different intensities applied. A maximum temperature change is observed at the Curie point achieving 2°C. Similar laws are also valid for other ferromagnetics with the same order of temperature change (1–2°C). For paramagnetics, whose magnetization obeys the Curie law $M \sim 1/T$, the magnetocaloric effect increases with lowering temperature.

Note that adiabatic demagnetization of paramagnetics is used in order to obtain temperatures close to absolute zero.

For MFs with the properties cited in Table 2.2 and $-(dM/dT)_{\rho,H} = 10^2$ A/(m · K) and field stress change $\Delta H \sim 10^6$ A/m formula (2.6) gives a temperature change of about 0.02 K.

2.4 CLOSED SYSTEM OF EQUATIONS

The equations of motion for stressed liquids have the form

$$\rho(\partial v_i/\partial t + v_k \partial v_i/\partial x_k) = \partial \sigma_{ik}/\partial x_k \tag{2.26}$$

The stress tensor in a MF is determined by relation (2.23). Viscous stresses are specified by the Newton law of viscous friction

$$\sigma'_{ik} = \eta\left(\frac{\partial v_i}{\partial x_k} + \frac{\partial v_k}{\partial x_i}\right) + \left(\eta_v - \frac{2}{3}\eta\right)\frac{\partial v_l}{\partial x_l}\delta_{ik} \tag{2.27}$$

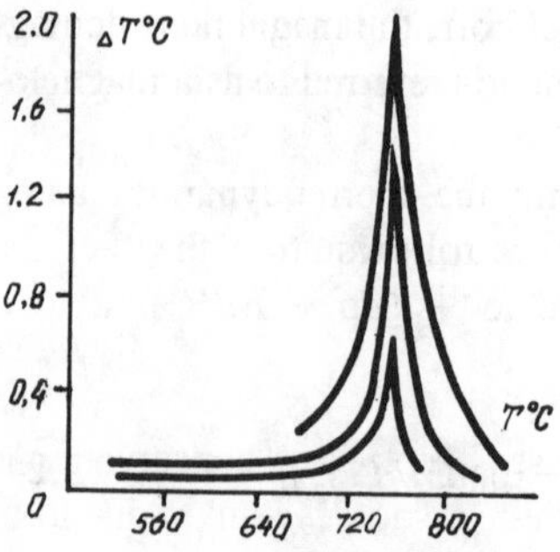

FIG. 2.4

As the size of ferromagnetic particles in a magnetic fluid increases, the non-Newtonian properties of the fluid become ever more prominent in ferromagnetic suspensions [44]. First of all, it is true of the yield limit stipulated by structurization of the solid phase. All this gives grounds for applying non-Newtonian viscous stress tensors to the MF, which, however, is not the subject of our book.

Substitution of (2.23), (2.27) into (2.26) with regard for the parallelism of vectors $\vec{M}$, $\vec{H}$, $\vec{B}$ gives the following equation of motion for a magnetic fluid

$$\rho\left(\frac{\partial v_i}{\partial t} + v_k \frac{\partial v_i}{\partial x_k}\right) = -\frac{\partial P}{\partial x_i} + \frac{\partial}{\partial x_k}\left[\eta\left(\frac{\partial v_i}{\partial x_k} + \frac{\partial v_k}{\partial x_i}\right)\right] + \frac{\partial}{\partial x_i}\left[\left(\eta_v - \frac{2}{3}\eta\right)\frac{\partial v_l}{\partial x_l}\right] + \rho g_i + [\mathrm{rot}\,\vec{H} \times \vec{B}]_i + \mu_0 M \frac{\partial H}{\partial x_i} \quad (2.28)$$

The i-th projection of the gravitational force ρg_i is introduced into the rhs of this equation.

The special nonuniformity of the viscosity coefficient η in a magnetic fluid may be stipulated by inhomogeneity of the magnetic field which it depends on, generally speaking.

The last two terms in the rhs of the equation of motion stand for the fluid magnetic field interaction. The former specifies the force of interaction of the conduction currents and the field, being the main one in classical magnetic hydrodynamics. So far as the motion of charges in a fluid is required for this force to be manifested, it may be called magnetodynamic. Another force $\mu_0 M \nabla H$ is determined by the interaction of the continuum magnetic moment and inhomogeneous field. It may be termed a magnetostatic force and is the main one in the thermomechanics of magnetic fluids.

In the general case, both forces are of the same order of magnitude and absolutely equitable in the broad sense of magnetic hydrodynamics.

For comparison, let us estimate all the three volume forces included in the equation of motion (2.28). With fluid density $\rho \sim 10^3$ kg/m^3 and $g \sim 10$ m/cm^2, the pressure gradient due to the gravitational force is 10^4 N/m^3. The magnetostatic force at $M = 50$ kA/m and $|\nabla H| = 10^6$ A/m^2 provides a fluid pressure gradient of 5×10^4 N/m^3. A pressure gradient of the same magnitude is caused by the magnetodynamic force in the field at induction $B = 1$ Tesla and a conduction current of 5 A/cm^2. The comparison shows that all the three forces can make an appreciable contribution to the hydrostatic pressure distribution in the fluid.

The continuity equation is of the form for an ordinary liquid

$$\partial\rho/\partial t + \mathrm{div}(\rho\,\vec{v}) = 0 \quad (2.29)$$

The equation for temperature, which expresses the law of conservation of energy in a magnetic fluid, is supplemented with heat sources (sinks) due to the magnetocaloric effect. Besides, it includes the terms for heat release due to viscous and Joulean energy dissipation and fluid compressibility

$$c_{\rho,H}(\partial T/\partial t + \vec{v}\nabla T) = \mathrm{div}(\lambda \nabla T) + \sigma'_{ik}(\partial v_i/\partial x_k) +$$
$$+ \sigma^{-1}(\mathrm{rot}\,\vec{H})^2 - T(\partial P/\partial T)_{\rho,H}\mathrm{div}\,\vec{v} - \tag{2.30}$$
$$- \mu_0 T(\partial M/\partial T)_{\rho,H}(\partial H/\partial t + \vec{v}\nabla H) + Q$$

The complete thermodynamic derivation of this equation is given in [12]. The magnetocaloric effect is associated both with the changing of magnetic field intensity with time $\partial H/\partial t$ and the fluid travel to regions with another magnetic field intensity $\vec{v}\ \nabla H$. The last term Q in the rhs of equation (2.30) describes other possible heat sources in the fluid.

Maxwell equations for an electromagnetic field in continuum at conventional assumptions in magnetic hydrodynamics [8] are written as follows

$$\left.\begin{aligned} &\partial \vec{B}/\partial t = \mathrm{rot}[\vec{v}\times\vec{B}] - \frac{1}{\sigma}\mathrm{rot}[\mathrm{rot}\,\vec{H}];\ \mathrm{div}\,\vec{B} = 0; \\ &\mathrm{rot}\,\vec{H} = \vec{j};\ \vec{j} = \sigma[\vec{E} + \vec{v}\times\vec{B}] \end{aligned}\right\} \tag{2.31}$$

In these equations, an important generalization for a magnetic fluid is the difference between the induction $\vec{B}$ and intensity $\vec{H}$ of a magnetic field in continuum. These vectors are related via fluid magnetization

$$\vec{B} = \mu_0(\vec{H} + \vec{M}), \vec{M} = (M/H)\,\vec{H} \tag{2.32}$$

Taking account of the parallelism of these vectors $\vec{B} = \mu_0(1 + M/H)\vec{H} = \mu_0(1 + \chi_s)\vec{H} = \mu_s H$, $\mu_s = \mu_0(1 + \chi_s)$, $\chi_s = M/H$. Both magnetic susceptibility χ_s and magnetic permeability μ_s of the fluid alike are the functions of magnetic field intensity.

Equations of state close the system of thermomechanic equations for magnetic fluids and include the law of magnetization:

$$\rho = \rho(P,T),\ M = M(\rho,T,H) \tag{2.33}$$

It should be noted that the fact that magnetic susceptibility of fluid ($\chi_s \to 0$) tends to zero or magnetic permeability ($\mu \to \mu_0$) tends to μ_0 need not necessarily point to disappearing magnetic properties of the fluid and may consist with a state of its magnetic saturation $M = M_s$ = const at sufficiently high magnetic field intensities. Thus, it is of several hundredths in the fields of 10^6 A/m χ_s (see Fig. 2.2).

At this stage, the most widely used and studied is the model of nonconducting ($\sigma = 0$) incompressible magnetic fluid with constant transport coefficients (η = const, λ = const); with neglect of all heat sources in the equation for temperature, and with linear equations of state. The following system of equations is consistent with this model:

$$\left.\begin{array}{l} \rho[\partial\vec{v}/\partial t + (\vec{v}\nabla)\vec{v}] = -\nabla P + \eta\Delta\vec{v} + \mu_0 M\nabla H + \rho\vec{g}; \\ \operatorname{div}\vec{v} = 0; \\ \partial T/\partial t + \vec{v}\nabla T = \kappa\Delta T; \\ \operatorname{rot}\vec{H} = 0;\ \operatorname{div}(\vec{H} + \vec{M}) = 0;\ \vec{M} = (M/H)\vec{H}; \\ \rho = \rho^*[1 - \beta_\rho(T - T^*)];\ M = M^* - K(T - T^*) + \chi_r(H - H^*) \end{array}\right\} \quad (2.34)$$

At small magnetic field intensities corresponding to the initial section of the magnetization curve, the law of magnetization may be used in the form

$$\vec{M} = \chi\vec{H},\ \chi = \chi_r = \chi_s = \text{const}$$

The stress tensor for this model is

$$\sigma_{ik} = -P\delta_{ik} + \eta(\partial v_i/\partial x_k + \partial v_k/\partial x_i) - (1/2)\mu_0 H^2\delta_{ik} + H_i B_k \quad (2.35)$$

2.5 BOUNDARY CONDITIONS. MAGNETIC PRESSURE JUMP

For fluid velocity, the boundary conditions, as usually, imply the adhesion of a fluid to a solid surface $\vec{v} = 0$. For a nonviscous fluid, suffice it to know the condition when the fluid cannot flow through the solid surface ($\vec{v} \cdot \vec{n} = 0$, where $\vec{n}$ is the unit vector of the normal to the surface).

At the interface of two viscous fluids, when the molecular interaction is known to exclude their slipping, the tangential velocity components must be equal $\vec{v}_{1\tau} = \vec{v}_{2\tau}$.

At the deformable interface, the kinematic condition must hold which says that the boundary velocity be equal to the normal velocity component of the fluid on it $\partial\xi/\partial t = (\vec{v}\vec{n})$.

If the surface, for instance, is prescribed in the form $z = f[x(t), y(t), t]$, then the kinematic condition on it may be written as

$$\partial f/\partial t + \vec{v}\nabla f = v_z, \text{ where } \vec{v}\nabla f = v_x\partial f/\partial x + v_y\,\partial f/\partial y$$

For temperature *T*, the boundary conditions are to prescribe its value or heat flux density $\vec{q} = -\lambda\nabla T$ on the confining walls. For convective heat transfer between the media, the heat flux density $\overline{q}$ at the interface may be prescribed by the Newton law

$$\vec{q}\,\vec{n} = a_T(T - T_c) \tag{2.36}$$

where a_1 is the coefficient of convective heat transfer from the surface; T_0 is the temperature of a liquid or gas flow around the surface.

The conjunction of thermal conditions at the interface between media 1 and 2 implies that heat flux temperatures and densities at this boundary be prescribed equal:

$$T_1 = T_2, \vec{q}_1 \cdot \vec{n} = \vec{q}_2 \cdot \vec{n} \tag{2.37}$$

Boundary conditions for a magnetic field include continuity of the normal induction component B_n when passing through the interface of media with different magnetization

$$B_{1n} = B_{2n} \text{ or } (\vec{B}_1 - \vec{B}_2) \cdot \vec{n} = 0 \tag{2.38}$$

This condition determines a jump in the normal field intensity component H_n, equal to the difference of normal fluid magnetization components

$$H_{1n} - H_{2n} = M_{2n} - M_{1n} \tag{2.39}$$

Still another condition implies that the tangential field intensity component $\vec{H}_\tau$ in conducting media undergoes a sudden change specified by the density of surface current $\vec{i}$ at the interface:

$$(\vec{H}_1 - \vec{H}_2) \times \vec{n} = \vec{i} \tag{2.40}$$

The normal vector $\vec{n}$ is directed into medium 2. In nonconducting media $\vec{i} = 0$ and the conjugation condition for a magnetic field involves continuity of the normal induction and tangential field intensity components when passing through the interface:

$$(\vec{B}_1 - \vec{B}_2)\vec{n} = 0; (\vec{H}_1 - \vec{H}_2) \times \vec{n} = 0; B_{1n} = B_{2n}, \vec{H}_{1\tau} = \vec{H}_{2\tau} \tag{2.41}$$

Such conjugation boundary conditions greatly hinder the solution of the problems, particularly when the shape of the interface is unknown, as is often the case in dealing with magnetic fluids.

The boundary conditions of the first kind are more convenient when the values of sought—for functions at the interface are to be prescribed. These situations are possible for a magnetic field when one of the boundary media has high magnetic permeability. This is easy to understand when we consider the refraction of magnetic flux lines at the interface of two media. At this interface we have (2.41) $B_{1n} = B_{2n}$, $H_{1\tau} = H_{2\tau}$ or with allowance for $\vec{B} = \mu\vec{H}$ $B_{1\tau}/B_{2\tau} = \mu_1/\mu_2$. By introducing the angle α between the magnetix flux line and the normal to the interface so that tg $\alpha = B_\tau/B_n$, we obtain the law of refraction of the magnetic flux lines

$$\mathrm{tg}\,\alpha_1/\mathrm{tg}\,\alpha_2 = \mu_1/\mu_2 \tag{2.42}$$

If, for example, medium 2 has high magnetic permeability ($\mu_2/\mu_0 \to \infty$), then this will mean tg $\alpha_1 = 0$ or $\alpha_1 = 0$ for medium 1. Therefore, the magnetic field flux lines enter the medium of infinitely large magnetic permeability at right angles. Thus, for a magnetic field in the medium having small magnetic permeability ($\mu/\mu_0 \sim 1$), the boundary condition of zero tangential induction component and, hence, of field intensity ($B_{1\tau} = H_{1\tau} = 0$) may be set at its interface with the medium with $\mu/\mu_0 \to \infty$. This condition is of meaning only when the magnetic field lines intersect this boundary. There may be situations when this condition is not fulfilled, i.e., when the magnetic field is primarily tangential at all interface points. The latter, for instance, takes place if the line linear conductor is surrounded by a coaxial annular cylindrical layer of a highly magnetic substance. In this case, the field lines do not intersect any of the interfaces; they are not refracted, and the magnetic field intensity at all points in space is the same as in vacuo, i.e., in the absence of a magnetic substance. Naturally, the field flux density in the magnetic layer is μ/μ_0 as large as that in vacuo.

Another condition that stems from (2.42) at $\mu_2/\mu_0 \to \infty$ is that it may mean tg $\alpha_2 = \infty$ or $\alpha_2 = \pi/2$, as viewed from medium 2, i.e., in the medium with infinitely large magnetic permeability the magnetic field lines are tangential to the interface with a nonmagnet. At the interface inside a highly magnetized substance, the normal field flux density and intensity components are zero ($B_{2n} = H_{2n} = 0$). This happens because practically all the magnetic field flux lines are concentrated inside a high magnet. These situations are illustrated by Fig. 2.5.

Dynamic boundary conditions are most important in the consideration of the phenomena that occur at the liquid-liquid, liquid-gas interfaces. In the most general form they relate the balance of stresses at the interface that ensures thermodynamic equilibrium of the system. Neglecting the surface tension, these conditions reduce to give the equality of stresses acting on both sides of the interface:

$$(\sigma_{2ik} - \sigma_{1ik})n_k = 0 \tag{2.43}$$

If surface tension forces are being considered, the rhs of (2.43) is supplemented with the capillary pressure normal to the interface and determined by its curvature:

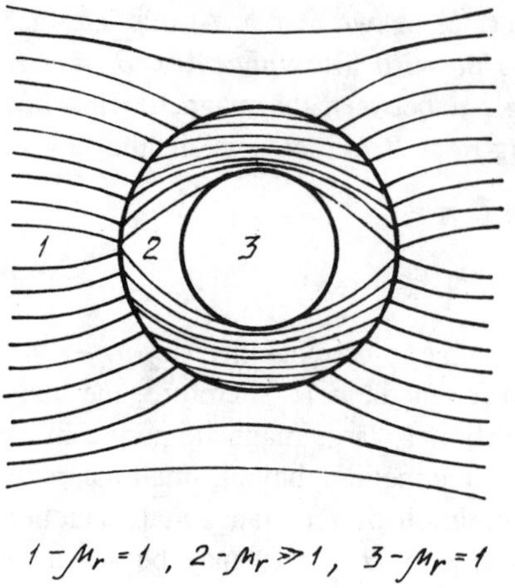

FIG. 2.5

$$(\sigma_{2ik} - \sigma_{1ik})n_k = \alpha(1/R_1 + 1/R_2)\, n_i \tag{2.44}$$

The main surface curvature radii R_1 and R_2 here are considered positive if they are directed into medium 1.

Finally, situations are possible when the coefficient of surface tension along the surface appears to be variable. This may be caused, for example, by nonuniformity of the temperature this coefficient depends on or by inhomogeneity of various impurities. For liquids, the surface tension coefficient decreases with growing temperature, $\partial\alpha/\partial T < 0$, and in a first approximation its temperature dependence may be approximated by the linear function

$$\alpha = \alpha^* - \beta_\alpha(T - T^*);\ \beta_\alpha > 0 \tag{2.45}$$

The inhomogeneity of α is responsible for the situation when an extra tangential force $\vec{f}_\tau$ starts acting at the interface and tends to set the liquid surface in motion in the direction when the surface tension coefficient $\vec{f}_\tau$ = grad α increases. So far as α is only determined at the interface, its gradient is also estimated at this very interface and is tangential to it at each point. In virtue of this, $(\text{grad}\ \alpha)_i\ n_i = 0$.

Taing account of the latter factor, the dynamic boundary conditions at the liquid interface are finally written as [2]

$$(\sigma_{2ik} - \sigma_{1ik})n_k = \alpha(1/R_1 + 1/R_2)n_i - (\text{grad}\ \alpha)_i \tag{2.46}$$

(The normal vector $\vec{n}$ is directed to medium 2).

The stress tensor in a magnetic fluid is determined by expression (2.23). Of special interest are the relations for the Maxwellian stresses when passing through the interface of nonconducting media $(\sigma^e_{2ik} - \sigma^e_{1ik})n_k)$. In this connection, consider the difference of shear $\sigma^e_{\tau n}$ and normal σ^e_{nn} stresses. Then $\sigma^e_{2\tau n}$ −

$\sigma^e_{1\tau n} = H_{2\tau}B_{2n} - H_{1\tau}B_{1n} = 0$ with allowance for boundary conditions (2.41) for a magnetic field. Simple calculations may show that

$$\sigma^e_{2nn} - \sigma^e_{1nn} = (1/2)\mu_0(H_1^2 - H_2^2) + H_{2n}B_{2n} - H_{1n}B_{1n}$$
$$= (1/2)\mu_0[(\vec{M}_1\vec{n})^2 - (\vec{M}_2\vec{n})^2]$$

It follows from the relationships obtained that the tangential components of the Maxwellian stress tensors are continuous when passing through the interface of nonconducting magnetic media while the normal component is discontinuous in proportion to the difference of squared normal magnetization components of the media. This fact is presented mathematically as follows:

$$(\sigma^e_{2ik} - \sigma^e_{1ik})n_k = (1/2)\mu_0[(\vec{M}_1\vec{n})^2 - (\vec{M}_2\vec{n})^2]n_i \qquad (2.47)$$

Substituting the explicit form of stress tensor (2.23) into (2.46) and taking account of (2.47), we may write the dynamic boundary conditions as

$$\{P_1 - P_2 - \alpha(1/R_1 + 1/R_2) + (1/2)\mu_0[(\vec{M}_1\vec{n})^2 - (\vec{M}_2\vec{n})^2]\}n_i = \\ = (\sigma'_{1ik} - \sigma'_{2ik})n_k - (\text{grad}\,\alpha)_i \qquad (2.48)$$

Magnetic Pressure Jump

The projection of expression (2.48) onto the normal to the surface determines, in particular, the pressure jump at the interface of stationary media

$$P_1 - P_2 = \alpha(1/R_1 + 1/R_2) - (1/2)\mu_0[(\vec{M}_1\vec{n})^2 - (\vec{M}_2\vec{n})^2] \qquad (2.49)$$

As is seen from (2.49), this pressure jump is indebted to two factors. Just as in an ordinary liquid the first is determined by capillary forces on a curved surface and causes a capillary pressure jump to occur. The second factor is the difference in the magnetic characteristics of liquids responsible for the magnetic pressure jumps and proportional to the difference of squared normal magnetization components of the media. The magnetic pressure jump is independent of the surface curvature and occurs at a plane interface. Due to this jump, pressure in a liquid of higher magnetization is lower. In particular, when the magnetic fluid is in contact with atmosphere, the pressure inside the magnetic fluid at the plane interface is less than the atmospheric one by $(1/2)\,\mu_0 M_n^2$. The physical nature of the magnetic pressure jump can easily be understood should we recall the discontinuity of the normal magnetic field intensity component $H_{2n} - H_{1n} = M_{1n} - M_{2n}$ that occurs at the interface of media with different magnetization. Physically, this change of the field intensity proceeds in thin surface liquid layers of the order of molecular thickness δ (Fig. 2.6). Hence, the surface molecules of contacting liquids are really in a nonuniform magnetic field whose gra-

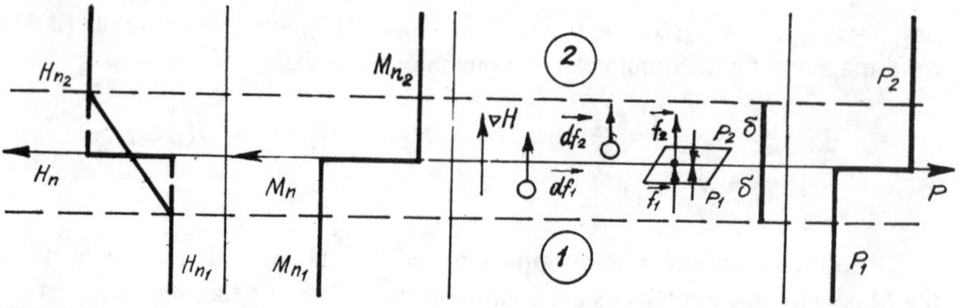

FIG. 2.6

dient is directed into the medium of lower magnetization (field intensity in this medium is higher). The magnetic force is acting upon them in the same direction, thus increasing the pressure in the medium of lower magnetization and decreasing it in the medium of higher magnetization. A magnetic pressure jump occurs at the interface. This jump may be determined numerically as well.

In the surface layer of every liquid the field changes by $(H_{2n} - H_{1n})/2$ and its gradient is of the order of $(H_{2n} - H_{1n})/2\delta$. Each magnetic particle with a moment $\vec{m}$ is affected by a magnetic force $\vec{df} = \mu_0(\vec{m}\nabla)\vec{H}$. As the tangential field component is continuous, and, when passing through the boundary, changing is only the normal field component, the force $\vec{df}$, determined by the derivatives of appropriate field components, has only the component $df_n = \mu_0(\vec{m}\nabla)H_n$ which is normal to the surface. It is also evident that the field only changes along the normal to the surface: therefore, the operator $(\vec{m}\nabla)$ includes only the normal derivatives $m_n\nabla_n$. Then $df_n = \mu_0 m_n \nabla_n H_n$. As already noted, $\nabla_n H_n = (H_{2n} - H_{1n})/2\delta$ and, thus, $df_n = \mu_0 m_n (H_{2n} - H_{1n})2\delta$. To be certain, assume $H_{2n} > H_{1n}$, i.e., $M_{2h} < M_{1h}$. In the surface fluid layer of thickness δ as well as on the unit area there are $n\delta$ magnetic particles (n being their number in a unit volume). Therefore, from the side of each fluid a unit interface will be affected by a force $f_n = df_n \cdot n\delta = \mu_0 m_n\, n(H_{2n} - H_{1n})/2$. The product $m_n n$ is none other than the sum of normal components of magnetic particles per unit volume, i.e., the normal magnetization component M_n of the fluid. Finally, $fn = \mu_0 M_n \times (H_{2n} - H_{1n})/2$. Taking account of the fact that a similar force acts upon the surface from the side of each fluid in one and the same direction (toward liquid 2 in the case), the condition of equilibrium for the flat interface between them will be written as $P_1 + f_{1n} + f_{2n} = P_2$ or

$$P_1 - P_2 = -(f_{1n} + f_{2n}) = -(1/2)\mu_0 (M_{1n} + M_{2n})(H_{2n} - H_{1n}) =$$
$$= -(1/2)\mu_0 (M_{1n} + M_{2n})(M_{1n} - M_{2n}) = -(1/2)\mu_0 (M_{1n}^2 - M_{2n}^2)$$

in accordance with (2.49).

If one of the fluids is nonmagnetic, then the above consideration shows that the layer particles of the magnetic fluid are affected by the outward force and responsible for the pressure loss in it by $(1/2)\,\mu_0 M_n^2$.

It should be noted in conclusion that of the greatest interest are the phenomena in which thermomechanic processes are interconnected with electromagnetic ones and depend on them, i.e., when the set of equations (2.34) is interrelated. First, the Maxwell and thermomechanics equations are related by the temperature dependence of fluid magnetization. Second, this relationship is very prominent, if the fluid has a free surface. In this case, the fluid motion greatly depends on the shape of this surface and, in turn, effects the latter. On the other hand, the shape of the surface is determined by the magnetic field configuration and affects its spatial distribution through boundary conditions (2.41). The interaction between all these factors results in new phenomena that are only inherent in magnetic fluids.

Therefore, of the greatest interest in magnetic fluids are the phenomena that proceed in isothermal conditions in the presence of a free fluid surface. Further chapters are mainly devoted to these very processes.

In view of the above, we omit such extended problems as isothermal fluid flows in closed channels with solid walls and flows about solids. In such situations, the fluid-field interaction results in simple re-estimation of fluid pressure which does not affect the fluid motion. The situation is just the other way round in dealing with nonequilibrium processes in a magnetic fluid which will be concerned with the concluding chapters of the monograph.

CHAPTER

THREE

STATICS OF MAGNETIC FLUIDS

Statics deals with phenomena that take place in a fluid remaining at rest. It also concerns the distribution of pressure, the floating of bodies and equilibrium shapes of the free surface of a fluid. For a magnetic fluid, these problems are of specific interest owing to the action of volume magnetic force $\mu_0 M \nabla H$.

3.1 MECHANICAL EQUILIBRIUM CONDITION. THERMOMAGNETIC CONVECTION MECHANISM

Potentiality of the volume force is a necessary condition for equilibrium in a fluid. As applied to a magnetic fluid in the gravitational or nonuniform magnetic fields, this means $\mathrm{rot}[\rho \vec{g} + \mu_0 M \nabla H] = 0$. From this condition and with account taken of the equations of state it follows

$$[\beta_\rho \rho \vec{g} + \mu_0 K \nabla H] \times \nabla T = 0 \tag{3.1}$$

It is evident that mechanical equilibrium is possible in an isothermal fluid, $\nabla T = 0$.

Equation (3.1) also implies that the fluid may be at rest if the acceleration of gravity $\vec{g}$, field gradients ∇H and temperature gradients ∇T lie on the same straight line, i.e. they are colinear and, in a general case, if $\nabla T \| (\beta_\rho \rho \vec{g} + \mu_0 K \nabla H)$. A particular case when the latter condition is realized is the fulfilment of the condition $\nabla H = -(\beta_\rho \rho / \mu_0 K)\vec{g}$. This is the case of complete mutual neutralization of the buoyancy and magnetic forces.

If these conditions are not fulfilled, then convective motion in the fluid is inevitable. In normal liquids, this motion is induced by the buoyancy force that occurs in a nonuniformly heated liquid, to be more specific, in a liquid with nonuniform density contribution. In this case, the lighter layers go upward and the heavier ones go downward.

In a magnetic fluid, this mechanism works together with a magnetic mechanism. Its physics can easily be understood through the situation when the magnetic fluid intensity gradient coincides in its direction with the temperature gradient. Then, in a state of mechanical equilibrium, a static pressure gradient $\nabla P = \mu_0 M \nabla H$ appears in a magnetic fluid. Let us consider the forces acting upon a fluid element when it is shifted from the equilibrium position with its temperature T_0 and magnetization M_0. As this element is shifted towards the temperature gradient, it reaches the region of higher temperature $T_1 > T_0$ and lower magnetization $M_1 < M_0$ (it is assumed that $dM/dT < 0$). When in this position it is affected in different directions by the magnetic force $\mu_0 M_0 \nabla H$ and pressure gradient $\nabla P_1 = \mu_0 M_1 \nabla H$. The magnetic force is predominant ($M_0 > M_1$) and the resultant force is directed to the field gradient, i.e., from the position of equilibrium (since $\nabla H \upuparrows \nabla T$).

If the fluid element under consideration moves in the direction opposite to the temperature gradient (to the cold region), then the resultant force is determined by the pressure gradient and is directed to the region of lower pressures, i.e., opposite to ∇H, and, hence, to ∇T, again from the position of equilibrium. Thus, in case of parallel vectors ∇H and ∇T, any displacement of the fluid element from the position of equilibrium brings about the forces which make it nonequilibrium and promote convective motion in the fluid. Just like in a normal heavy liquid where more dense layers tend to move down if less dense layers are below them, in a magnetic fluid the layers with higher magnetization tend to move in the direction of the magnetic fluid intensity gradient provided the layers of lower magnetization lie before them. Such an unstable stratification may be caused not only by the temperature gradient but also by the admixture concentration gradient.

No thermoconvective motion will be observed if more dense layers in a normal liquid are below, and magnetization in a magnetic fluid is increasing in the field gradient direction when the temperature gradient is opposite to the field gradient in direction.

Naturally, all the above will be exactly to the contrary in fluids with anomalous dependences $\rho(T)$ and $M(T)$ ($\partial\rho/\partial T > 0$, $\partial M/\partial T > 0$).

The above consideration also shows that even when the condition of possible mechanical equilibrium (3.1) is fulfilled ($\vec{g} \| \nabla H \| \nabla T$), this does not guarantee that the fluid does not move, i.e., equilibrium may turn out to be unstable. It is stable when ∇T is opposite in direction to $\vec{g}$ and $\overrightarrow{\nabla H}$ ($\nabla T \updownarrows \vec{g} \downdownarrows \nabla H$). It is unstable when all the three vectors are parallel ($\nabla T \downdownarrows \vec{g} \downdownarrows \nabla H$). In other situations, the stability of equilibrium is determined by the competing action of buoyancy and thermomagnetic forces. This problem will be discussed in greater detail in subsequent chapters.

In the situations when free liquid surfaces are concerned, the condition of possible mechanical equilibrium ought to be supplemented by the condition of constancy of the surface tension coefficient grad $\alpha = 0$ or, in particular, by the condition of constancy of the surface temperature. In case isothermicity of the liquid interface is disturbed, thermocapillary convection appears in liquids which will also be a subject of further discussion.

3.2 PRESSURE DISTRIBUTION AND FLOATING OF BODIES IN MAGNETIC FLUID

Pressure distribution in a stationary magnetic fluid is described by the static equilibrium equation

$$\nabla P = \rho \vec{g} + \mu_0 M \nabla H \tag{3.2}$$

which implies that the pressure gradient due to magnetostatic force is parallel to the magnetic field intensity gradient. In fluid, pressure is higher at greater magnetic field intensities.

In an isothermal fluid, the general integral of this equation is of the form

$$P = P_0 - \rho g (z - z_0) + \mu_0 \int_{H_0}^{H} M dH \tag{3.3}$$

where P_0 is pressure at a point (x_0, y_0, z_0) where $H = H_0$; the axis z is directed vertically upward. In (3.3), the integral is taken with allowance for the magnetization equation (for instance, (2.4)) with the known space distribution of the field $H = H(x,y,z)$ emanating from the solution of the Maxwell equation.

Analysis of equation (3.2) shows that magnetostatic force may increase the gravitational pressure gradient (at $\nabla H \downarrow\downarrow \vec{g}$) and as well as decrease it (at $\nabla H \uparrow\downarrow g$) thus ensuring, in particular, its total neutralization, i.e., realizing zero gravity in the fluid. At $\nabla H = -(\rho/\mu_0 M)\vec{g}$, we have $\nabla P = 0$ and $P =$ const. A similar effect may also be achieved in conductive fluid by virtue of magnetodynamic force $[\vec{j} \times \vec{B}]$, which, however, requires that some energy be expended for the passage of current. In a magnetic fluid, this expenditure of energy is totally unnecessary when a magnetic field is set up by permanent magnets.

It is evident from the above that the floating of bodies in a MF (magnetic fluid) may be effectively controlled.

Floating of Nonmagnetic Bodies in MF

There are two forces acting upon a nonmagnetic body immersed in a magnetic fluid. One of them is the gravitational force and the other is due to the pressure gradient and is opposite to it in direction. In the simplest situation, when magnetization of the fluid may be considered constant, $M =$ const, the magnetic field gradient is directed vertically downward and may also have a constant value G, $\nabla H = [0,0\text{–}G]$. From (3.3) we have

$$P = P_0 - (\rho + \mu_0 MG/g)g(z - z_0) = P_0 - \rho_{\text{eff}} g(z - z_0)$$

In such a situation, the behavior of bodies in the fluid is equivalent to their behavior in a normal liquid at a density $\rho_{\text{eff}} = \rho + \mu_0 MG/g$. Bodies having a density ρ_T, ρ_{eff} will sink in the fluid; those having $\rho_T < \rho_{\text{eff}}$ will float up. Thus, the floating-up condition of bodies may be written as

$$\rho_T < \rho + \mu_0 MG/g \tag{3.4}$$

in contrast to the condition $\rho_T < \rho$ for a normal liquid. It is obvious that floating-up condition (3.4) may also be fulfilled at a body density higher than that of the fluid $\rho_T > \rho$.

Magnetostatic interaction between a fluid and a nonuniform magnetic field provides an additional expulsive force which enables the bodies having a density higher than that of the magnetic fluid to float up. When the magnetic field intensity gradient is opposite in direction to the gravitational force ($G < 0$), the magnetostatic force reduces the expulsive buoyancy force. As a result, bodies having a density less than that of the magnetic fluid will sink. Provided $\mu_0 MG = -\rho g$, the gravitational force in the MF is completely neutralized by the magnetic one, and the bodies having any small density will sink in it.

In the general case, the force the MF exerts on the body is determined by a flux of momentum through its surface $\vec{F}_T = \oint_S \sigma_{nn} \vec{n}\, dS$, i.e., by the integral of normal stresses in the fluid over its surface S. For a nonmagnetic body this expression is transformed to the form

$$\vec{F}_T = \oint_S [P + (1/2)\mu_0 (\vec{M}\vec{n})^2]\, \vec{n}\, dS \tag{3.5}$$

taking into account that the flux of the Maxwell stress tensor from inside of the body through its surface is zero, i.e., the nonmagnetic body is not affected by a volume magnetic force. The pressure distribution P in the fluid is specified by expression (3.3); $\vec{n}$ is the outward normal to the body surface.

As a nonmagnetic body is introduced into the fluid, the magnetic field in it is destorted. This distortion is of the order of fluid magnetization M magnitude. In the widely spread case when these distortions may be neglected ($M \ll H$), the force causing the bodies immersed in the MF to move is related as

$$\vec{F}_T = -\oint_S P\vec{n}\, dS = \oint_S [\rho g z - \mu_0 \int_{H_0}^{H} M dH]\, \vec{n}\, dS \tag{3.6}$$

In this approximation, the magnetic field and fluid magnetization distributions are assumed to be the same before the body was introduced into it. Then the Ostrogradsky theorem may be applied to surface integrals (3.6) to transform them into the volume integrals over the volume V occupied by the body, bearing in mind that the integrand functions inside this volume have the value equal to the one they had without the body:

$$\vec{F}_T = -\oint_S P\vec{n}\,dS = -\int_V \nabla P\,dV = -\int_V (\rho\vec{g} + \mu_0 M\nabla H)\,dV =$$
$$= -(\vec{F}_g + \vec{F}_M) \tag{3.7}$$

Proceeding from this, the known Archimedes' principle is amended for a magnetic fluid to be: "A nonmagnetic body immersed in a magnetic fluid is affected by an expulsive force equal to the resultant force of the gravitational and magnetic weights of the fluid expelled by this body. The direction of the expulsive force is opposite to this resultant force."

The gravitational weight of the fluid volume $V\ \vec{F}_g = \int_V \rho\vec{g}\,dV = \rho\vec{g}V$, and the magnetic weight is understood to be the force acting upon a MF volume V in a nonuniform magnetic field $\vec{F}_M = \int_V \mu_0 M\nabla H dV$. If the fluid magnetization and the field gradient may be considered constant, within the fluid volume V expelled by a body, then from (3.7) it follows

$$\vec{F}_T = -(\rho\vec{g} + \mu_0 M\nabla H)V \tag{3.8}$$

Taking account of the gravitational force $\rho_T\vec{g}\,V$, acting upon the body, the total force which determines its motion in the fluid is

$$\vec{F} = [(\rho_T - \rho)\vec{g} - \mu_0 M\nabla H]V \tag{3.9}$$

whence, in particular, follow the conditions of the body's floating-up (3.4).

The horizontal components of the magnetic field gradient yield horizontal projections of the expulsive force and are able to ensure horizontal motion of bodies in the fluid from high-pressure to low-pressure regions, i.e., from the regions of high field intensity to the regions of low field intensity.

This effect enables one to implement density separation of solid mixtures. The possibility of carrying out such production process has been known for long. In view of the above, the advantages of the magnetostatic separator are evident.

The sufficient simplicity of establishing magnetic fields of different spatial configuration with the aid of permanent magnets makes it possible to realize various static pressure distributions in MFs. For instance, the formation in the fluid of a point from which the field intensity starts increasing in all the three directions provides a three-dimensional pressure 'pit' wherein a nonmagnetic body can float, i.e., a three-dimensional suspension that may serve, for example, as an acceleration transducer (accelerometer).

The action of the magnetostatic expulsive force may be used in designing liquid bearings and shock-absorbers.

The retention of the magnetic fluid by the field in required places in the gaps between the moving parts of a mechanism underlies the design of magnetofluid seals and controlled lubrication systems.

Let us emphasize once again that all these processes may be implemented without energy consumption.

Floating of Magnetic Bodies

When a body that is readily magnetized, i.e., capable of acquiring its own magnetic moment per unit volume M_T is immersed in MF, then its unit volume is affected by a force $\mu_0 M_T \nabla H$ from the side of the magnetic field. Hence, formulae (3.4) specifying the body's floating-up conditions will naturally be correlated to give

$$(\rho_T - \rho)g < \mu_0(M - M_T)G \qquad (3.10)$$

Similarly, expression (3.9) is generalized to the total force acting upon the body

$$\vec{F} = [(\rho_T - \rho)\vec{g} + (M_T - M)\nabla H]V \qquad (3.11)$$

In this first approximation, the difference of fluid and body magnetizations in a nonuniform field plays the same part as does the difference of their densities in the gravity field. If the body magnetization is higher than the fluid one, the resultant magnetic force will be in the direction of the increasing field; otherwise, the resultant force is in the opposite direction. In the latter case, the fluid pressure gradient is predominant.

Distortions of the magnetic field due to the immersed body will, certainly, make the obtained relations more complex. Nevertheless, they may be applied for design calculations and quantitative estimations. The error they produce may be of the order of M/H.

Floating of permanent magnets seems to be one of the most attractive effects in MF statistics. When a magnetic field source (permanent magnet or current) is immersed in a magnetic fluid, then thanks to its own magnetic field alone it excites in the fluid the forces able to move it and, in particular, to keep it floating even in the presence of gravity. Of course, we should make a reservation, that this cannot take place in an infinite fluid volume, for this will contradict the laws of conservation. Here, the walls of a vessel separating the magnetic fluid from the surrounding (for example, nonmagnetic) medium are of great importance. These walls are an interface for the media having different magnetic characteristics and are responsible for such an asymmetry of the forces acting on the magnet at which their resultant force repels the magnet from the walls.

For instance (Fig. 3.1), when the bottom of the vessel with MF is an interface for magnetic (fluid) and nonmagnetic (air) media, then as the magnet approaches the bottom, the lines of force of its magnetic field condense above it and do not 'want' to enter the medium of lower magnetic permeability. This condensation increases the pressure under the magnet to create an expulsive force which can balance the gravity force. The same reasons prevent the magnet in the fluid from approaching the vessel walls, which, when in combination, provides its stable floating relative to its spatial travel in three directions.

In the experiments performed, the proper dish-type magnet weight being 6.2 g, the expulsive magnetic force acting on it was 15 g at a distance of 1.5 mm from the vessel wall.

The expulsive force acting upon the magnet is determined by the general expression

$$F_i = \oint_S \sigma_{ik} n_k dS = \oint_S (-Pn_i + \sigma^e_{ik} n_k) dS \tag{3.12}$$

For this expulsive force to be calculated, one should know the magnetic field configuration H, which is hardly possible in many cases.

Therefore, the situations allowing for an exact analytical solution, are all the more interesting. One of them will be discussed below. It is also peculiar of this situation that a current rather than a permanent magnet will serve as a source of magnetic field.

Let (Fig. 3.2) a horizontal thin plate be immersed in a MF and a current of linear density $\vec{i} = [0, i, 0]$ flows along the axis y in its plane. The current density is changed at regular intervals in its normal direction $i = i_0 \cos kx$. The plate with flowing current is l distant from the horizontal boundary separating the MF and nonmagnetic medium 3. Region 1 under the plate $(-l < z < 0)$ and unlimited region 2 above the plate $(z > 0)$ are filled with a magnetic fluid of susceptibility $\chi(\vec{M} = \chi\vec{H})$. The current in the plate may be considered as surface current, bearing in mind that the tangential component of the magnetic field intensity will be discontinued when the current passes through it. Therefore, the following boundary conditions for the field will hold at the interface

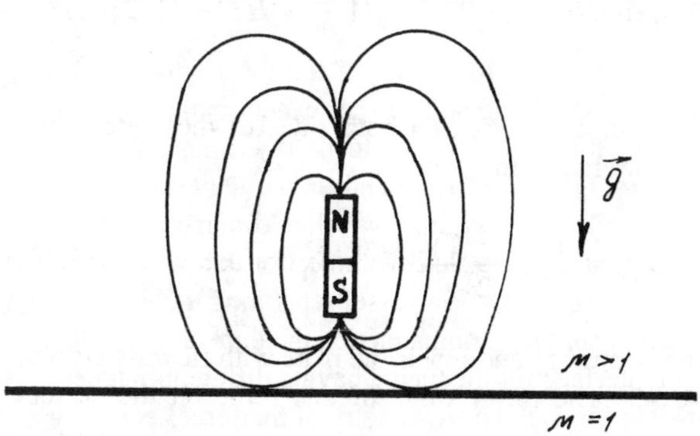

FIG. 3.1

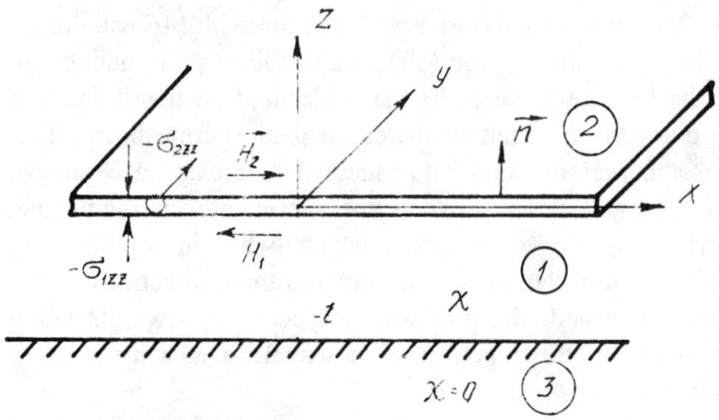

FIG. 3.2

$$
\begin{aligned}
z = 0: &\quad H_{2x} - H_{1x} = i; \; H_{2z} = H_{1z}; \\
z = -l: &\quad H_{1x} = H_{3x}, \; (1+\chi)H_{1z} = H_{3z}
\end{aligned}
\tag{3.13}
$$

In each of the regions, the magnetic field potentials $\Phi(\vec{H} = \nabla\Phi)$ satisfy the Laplace equation $\Delta\Phi = 0$ and, with regard to (3.13) and attenuation at infinity, have the form:

$$
\begin{aligned}
\Phi_1 &= -\frac{i_0}{2\kappa} e^{-\kappa l}\left[e^{\kappa(z+l)} + \frac{\chi}{2+\chi} e^{-\kappa(z+l)}\right]\sin\kappa x; \\
\Phi_2 &= \frac{i_0}{2\kappa}\left[e^{\kappa l} - \frac{\chi}{2+\chi} e^{-\kappa l}\right] e^{-\kappa(z+l)} \sin\kappa x; \\
\Phi_3 &= -\frac{i_0}{\kappa}\,\frac{1+\chi}{2+\chi} e^{\kappa z} \sin\kappa x
\end{aligned}
\tag{3.14}
$$

The normal stresses acting upon the plate from the fluid are specified as

$$
\sigma_{zz} = -P + H_z B_z - (1/2)\mu_0 H^2 = -P + \mu_0(1+\chi)H_z^2 - (1/2)\mu_0 H^2
\tag{3.15}
$$

The difference of these normal stresses on both sides of the plate is the vertical force P_2 acting per unit area of the plate:

$$
P_z = \sigma_{2zz} - \sigma_{1zz} = \frac{1}{2}\mu_0 \frac{\chi(1+\chi)}{2+\chi} e^{-2\kappa l} i_0^2 \cos^2\kappa x
\tag{3.16}
$$

This force is directed vertically upward and repels the plate with flowing current from the bottom of the vessel. The mean value of this force on the current change wavelength is

$$\overline{P}_z = \frac{1}{4}\mu_0 \frac{\chi(1+\chi)}{2+\chi} i_0^2 e^{-2\kappa l} \tag{3.17}$$

In this case, the magnetic expulsive force decreases exponentially with the growing distance to the vessel bottom. It attains its peak value at $l = 0$ to be $\overline{P}_z = \mu_0\chi(1 + \chi)i_0^2/4(2 + \chi)$. If its value is greater than that of the real gravitational force $P_g = (\rho_T - \rho)gh$ (h is the plate thickness), the plate will be floating at some distance from the bottom. This distance is found from the condition of equality of the above forces

$$l = \frac{1}{2\kappa}\ln\left[\frac{\mu_0\chi(1+\chi)i_0^2}{4(2+\chi)(\rho_T-\rho)gh}\right] \tag{3.18}$$

The floating of periodically magnetized magnetic bodies can accurately be calculated analytically in a similar way. In all these cases, the magnetic field potentials are estimated by exact solution of the Laplace equation of the form $e^{kz}\cos kx$.

As already noted, such consideration requires the knowledge of an exact magnetic field configuration, i.e., of those distortions of the magnetic field that are attributed to the interfaces of media. Nevertheless, we may easily overcome this difficulty by choosing a definite approach. It should be remembered that, following the third Newton's law, the force with which the fluid acts upon the magnet is equal and sign-opposite to the one with which the fluid is affected by the magnetic field. The latter is calculated by integrating the magnetic force density over the fluid volume:

$$\vec{F} = \int_V \mu_0 M\nabla H dV \tag{3.19}$$

and yields a non-zero resultant force even when the distribution of the magnetic field which a given source might have induced in an infinite medium is used. In such a situation, disturbances in the liquid distribution symmetry around the magnet due to vessel boundaries rather than magnetic field distortions play a decisive part. Thus in the considered case (cf. Fig. 3.2), the boundary $z = -l$ is responsible for the asymmetric position of the fluid above and below the plate. In other words, it creates above the plate an unbalanced fluid volume ($z > l$) whose attraction in the first approximation provides a vertical force acting upon the plate.

In such an interpretation, the force acting upon the magnetic field source is developed by the magnetic fluid volume provided by the boundaries of the vessel with the fluid and unbalanced as compared to the infinite medium.

This force can be calculated within an accuracy in the order, compared to unity, using the magnetic field configuration of the source in an infinite medium.

The magnetic field of the above plate with flowing current in an infinite medium ($l \to \infty$) is of the form:

$$\Phi_1 = -\frac{i_0}{2\kappa} e^{\kappa z} \sin \kappa x; \quad \Phi_2 = \frac{i_0}{2\kappa} e^{-\kappa z} \sin \kappa x \,;$$

$$H_2^2 = \left(\frac{\partial \Phi_2}{\partial x}\right)^2 + \left(\frac{\partial \Phi_2}{\partial z}\right)^2 = \frac{i_0^2}{4} e^{-2\kappa z}$$

This field acts upon an unbalanced fluid volume $z > l$ with a force opposite in direction to the axis z,

$$F_z = W \int_l^\infty (1/2)\mu_0 \chi \frac{\partial H_2^2}{\partial z} dz =$$

$$= \frac{1}{2}\mu_0 \chi [H_2^2(\infty) - H_2^2(l)] = -\frac{\mu_0 \chi i_0^2}{8} e^{-2\kappa l}\, W \tag{3.20}$$

(W is the fluid column area).

In terms of a unit surface F_z/W accurate in the χth order, the obtained expression coincides with (3.17) and has, as it should, an opposite sign. Certainly, in a majority of cases, an unbalanced fluid volume can hardly be defined as it was in the above consideration. Therefore, use should be made of general expression (3.19) for calculating the force. We shall emphasize once again that such an approach has an essential advantage, for it allows the configuration of a magnetic field in an infinite medium to be used in calculations. For magnets of simple shapes (cylindrical, spherical, elliptical) there are analytical expressions for the fields they induce.

3.3 EQULIBRIUM FORM OF THE SURFACE OF MAGNETIC FLUID AT REST

As far as mechanics is considered, the main property of the interface between two fluids is its ability to deform under the action of the forces applied to these fluids. The equilibrium form of such an interface is stipulated by the balance of normal stresses and capillary forces, i.e., by dynamic boundary conditions (2.46) and (2.49)

$$P_1 - P_2 = \alpha\left(\frac{1}{R_1} + \frac{1}{R_2}\right) - \frac{1}{2}\mu_0[(\vec{M}_1 \vec{n})^2 - (\vec{M}_2 \vec{n})^2] \tag{3.21}$$

Pressure distribution in each of the adjacent fluids 1 and 2 is specified by static equilibrium equation (3.2)

$$P = P_0 - \rho g(z - z_0) + \Phi \tag{3.22}$$

where $\Phi = \mu_0 \int_{H_0}^{H} M dH$ is the volume magnetic force potential.

Let the fluid interface be described by the equation $z = \xi(x,y)$. Then, substituting into (3.21) pressures (3.22) at the interface for each fluid, we arrive at the equation used to determine the shape of this interface

$$(\rho_1 - \rho_2)g\xi - \Phi_1 + \Phi_2 + \alpha\left(\frac{1}{R_1} + \frac{1}{R_2}\right) - \frac{1}{2}\mu_0[(\vec{M}_1\vec{n})^2 - (\vec{M}_2\vec{n})^2] = C, \tag{3.23}$$

where $\Phi_1 = \mu_0 \int_{H_0}^{H_2(\xi)} M_1 dH$; $\Phi_2 = \mu_0 \int_{H_0}^{H_2(\xi)} M_2 dH$

The constant C is found by prescribing pressure at some point of the space occupied by the fluids and fixes the pressure reference point. As the radii of curvature R_1, R_2 and the normal $\vec{n}$ are intricately nonlinearily expressed in terms of the first and second derivatives $\xi(x,y)$, differential equation (3.23) is rather difficult to solve. It is therefore reasonable to consider all its possible simplified versions.

If one of the media is a nonmagnetic gas, the interface will be referred to as a free fluid surface. This surface will be determined by the expression obtained from (3.23)

$$\rho g \xi - \Phi(\xi) + \alpha\left(\frac{1}{R_1} + \frac{1}{R_2}\right) - \frac{1}{2}\mu_0 (\vec{M}\vec{n})^2 = C \tag{3.24}$$

When the volume magnetic force is much greater than the other forces (gravitational, capillary, magnetic pressure jump), (3.24) implies $\Phi(\xi) = \text{const}$. Thus, without gravitational force, surface tension and magnetic pressure jump, the free fluid surface is the surface of the constant magnetostatic force potential. For a fluid magnetized to saturation ($M = \text{const.}$ $\Phi = \mu_0 MH + \text{const}$), this is the surface of constant magnetic field intensities $H(\xi) = \text{const}$. For the linear magnetization law ($M = \chi H$, $\Phi = (1/2)\mu_0\chi H^2 + \text{const}$), it is the surface of constant values of squared field intensity $H^2(\xi) = \text{const}$.

Taking account of the gravitational force yields the equation used to define the free surface shape

$$\xi = \Phi(\xi)/(\rho g) + \text{const} \tag{3.25}$$

Bearing in mind that magnetization is an increasing function of the magnetic field intensity, the values of the magnetostatic force potential Φ are always greater at higher field intensities. In the gravity field, therefore, the lift of the free MF surface will always be higher where the field intensities are higher. This fact may be turned to advantage in visualizing magnetic fields and in designing various display systems.

As a particular example of the solution of equation (3.25) consider the shape of a free MF surface in the field set up by a linear conductor. This situation is depicted in Fig. 3.3. The magnetic field of current 1 has an axial symmetry and only an asimuthal component $H = H\varphi = I/(2\pi\tau)$ everywhere tangential to the fluid surface, which excludes a magnetic pressure jump on it.

If the fluid is in a state of saturation magnetization (M = const), then the fluid surface rises at approaching the conductor, in accordance with the hyperbolic law $\xi = \frac{\mu_0 MI}{2\pi\rho g}\frac{1}{r} + C$. With the linear magnetization law, the surface shape is described by the square-law hyperbola $\xi = \frac{\mu_0 \chi I^2}{8\pi^2\rho g}\frac{1}{r^2} + C$. The value of the constant C may be assumed zero, which will correspond to the zero fluid level far from the current ($r \to \infty$).

The height of the fluid rise at the conductor may be estimated by taking $M = 5 \cdot 10^4$ A/m, $\rho = 10^3$ kg/m^3, $I = 10$ A and the conductor radius R = 1 mm. Then $\xi(R) \approx 100$ mm in the former case or at $\chi = 1\xi(R) = 0.2$ mm in the latter. As seen, with the linear law of magnetization, i.e., without additional magnetization of the fluid, its rise is very low because of an essential gravitational force. The effect of the latter may be drastically reduced provided a nonmagnetic fluid of similar density is above the MF surface. Then, ρ in the above formulae will be substituted by the difference of fluid densities Δρ which in a particular case may be equal to zero. This situation will be referred to as hydraulic zero gravity, since it is equivalent to zero gravity $g = 0$. In this case, the MF surface will be the surface of constant magnetic force potential Φ =

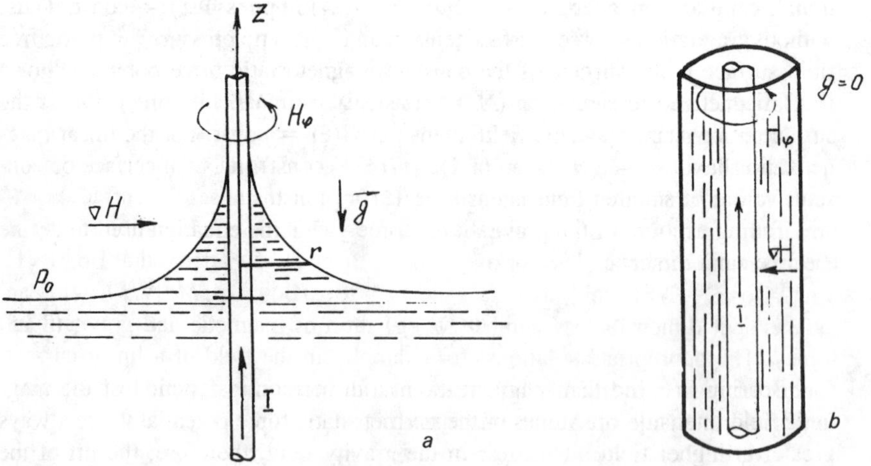

FIG. 3.3

const. which for the problem under consideration results in a surface equation of the form r = const. Thus, when in a state of hydraulic weightlessness, the MF is arranged as a cylindrical layer around a linear conductor with flowing current (Fig. 3.3b). The stability of such a layer related with capillary forces will be discussed below.

We shall only note here that, when describing the phenomena that take place on the MF surface, it is reasonable to specify the joint action of the magnetostatic $\mu_0 M|\nabla H|R$ and capillary a/R forces (R being the characteristic size of the system, for instance, the cylindrical column radius) in terms of a dimensionless criterion being the ratio of these forces $\mathrm{Bo}_m = \mu_0 M|\nabla H|R^2/\alpha$. This criterion may be referred to as the magnetic Bond number by analogy with the known gravitational Bond number $\mathrm{Bo} = \rho g R^2/\alpha$ which is a gravitational-to-capillary force ratio. If the magnetostatic force is predominant $Bo_m > 1$, then the above cylindrical layer is stable; otherwise ($\mathrm{Bo}_m < 1$), it disintegrates into droplets.

With the use of these criteria, Eq. (3.23) for the surface shape may be presented in a dimensionless form. Let us introduce a dimensionless surface coordinate $\tilde{\xi} = \xi/R$, dimensionless radii of curvature $\tilde{R}_1 = R_1/R$, $\tilde{R}_2 = R_2/R$, dimensionless magnetic force potentials $\tilde{\Phi} = \Phi/(\mu_0 M^*|\nabla H^*|R)$, and dimensionless magnetization $\vec{\tilde{M}} = \vec{M}/M^*$. The characteristic scale values R, M^*, $|\nabla H^*|$ are estimated from particular conditions of the problem. Equation (3.23) will then be written in a dimensionless form as

$$\mathrm{Bo}\,\tilde{\xi} - \mathrm{Bo}_m\,(\tilde{\Phi}_1 - \tilde{\Phi}_2) + (\tilde{R}_1^{-1} + \tilde{R}_2^{-1}) - S\,[(\vec{\tilde{M}}_1\vec{n})^2 - (\vec{\tilde{M}}_2\vec{n})^2] = \tilde{C} \tag{3.26}$$

where

$$\mathrm{Bo} = (\rho_1 - \rho_2)gR^2/\alpha, \quad \mathrm{Bo}_m = \mu_0 M^*|\nabla H^*|R^2/\alpha, \quad S = \mu_0 M^{*2} R/\alpha$$

The last of the criteria introduced, S, is the surface magnetic pressure-to-capillary jump ratio.

The above particular cases of Eq. (3.23) correspond to the predominant value of one or another dimensionless criterion. For instance, the surface equation in the form Φ = const complies with $\mathrm{Bo}_m \gg 1$, Bo, S. Equation (3.25) of the following dimensionless form $\tilde{\xi} = (Bo_m/\mathrm{Bo})\tilde{\Phi} + \tilde{C}$ is derived at $\mathrm{Bo}_m \gg 1$, S and $\mathrm{Bo} \gg 1$, S. In this case, the ratio $\Gamma = \mathrm{Bo}_m/\mathrm{Bo} = \mu_o M^*|\nabla H|/(\rho g)$ is an independent dimensionless number for the ratio of magnetic and gravitational forces. The appropriate solutions, for example, in the field of a linear current conductor may conveniently be written in a dimensionless form. In particular, for a fluid in a state of saturation magnetization

$$\tilde{\xi} = \Gamma/r$$

where $\Gamma = \mu_0 MI/(2\pi R^2 \rho g)$, $\tilde{\xi} = \xi/R$, $\tilde{r} = r/R$, R is the conductor radius. Here, the characteristic field intensity gradient is $|\nabla H^*| = I/(2\pi R^2)$. The numerical value of Γ yields, in this case, the maximum height of the fluid rise near the conductor in the units of its radius. At $\tilde{r} = 1$, we have $\tilde{\xi}_{max} = \Gamma$.

The above examples are a graphic evidence of the advantages due to the introduction of dimensionless parameters into differential equations and to the possibility of criterial interpretation of the results. From now on, we shall adhere to this approach when we find it possible and necessary.

Pinch Effect

A situation is known in magnetic hydrodynamics when a cylindrical plasma column is formed under the action of a line current passing through it, namely, linear pinch. It is caused by a magnetodynamic force that ensures self-contraction of the plasma. A similar situation is also possible in a conductive MF, with the only difference that the current passes through the fluid itself and the magnetic field inside the fluid is distributed as $H = H_\varphi = (1.2)jr$, where j is the current density. The magnetodynamic force to the column axis is related as

$$F_\partial = [\vec{j} \times \vec{B}]_r = -jB = -\mu_0(1 + \chi)jH = -(1/2)\mu_0(1 + \chi)j^2 r$$

The magnetic field intensity gradient in this case is directed from the column axis to the periphery. The magnetostatic force acts in the same direction

$$F_c = \mu_0 M(\partial H/\partial r) = (1/4)\mu_0 \chi j^2 r$$

preventing contraction of the column (Fig. 3.4). The resultant force is

$$F_p = -(1/4)\mu_0 j^2 r(2 + \chi)$$

At large values of χ ($\chi \gg 1$)

$$F_p = -(1/4)\mu_0 \chi j^2 r$$

and the magnetostatic force makes the pinch effect in a conductive MF twice as weak.

MF Droplets in a Magnetic Field

Of all possible configurations of fluids with a free surface of great interest are those which provide a closed surface. Such a volume will be referred to as a droplet.

For a fluid droplet subjected only to capillary forces, a spherical shape is most energy profitable. It is quite natural that the external uniform magnetic

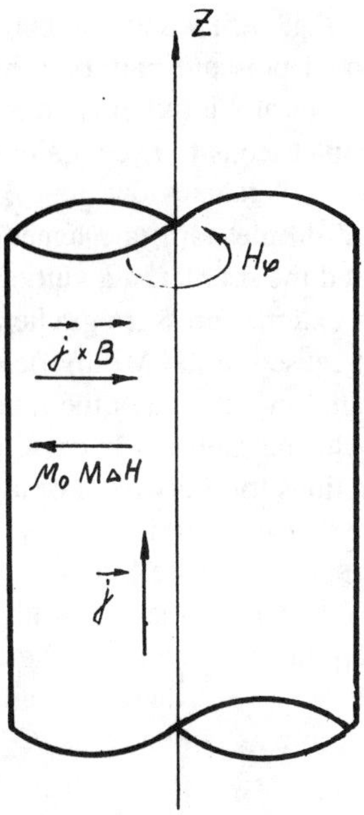

FIG. 3.4

field cannot displace the center of the MF droplet inertia. Nevertheless, together with capillary forces, it is significant in shaping the droplet configuration.

Consider a spherical MF droplet in a nonmagnetic media in a state of imponderability (Fig. 3.5). If the external magnetic field is uniform, then the field inside the droplet is also uniform and the volume magnetic force $\mu_0 M \nabla H$ is zero. However, a magnetic pressure jump takes place on the droplet surface. It is due to the magnetization vector component normal to the surface: $P_i - P_e = -(1/2)\mu_0 M_n^2$ (quantities determined inside the droplet have subscript i and those outside, subscript e). In accordance with this, a lower pressure will be observed inside the droplet as compared to the side points ($x = \pm R$) where the field component normal to the surface will be zero at the upper

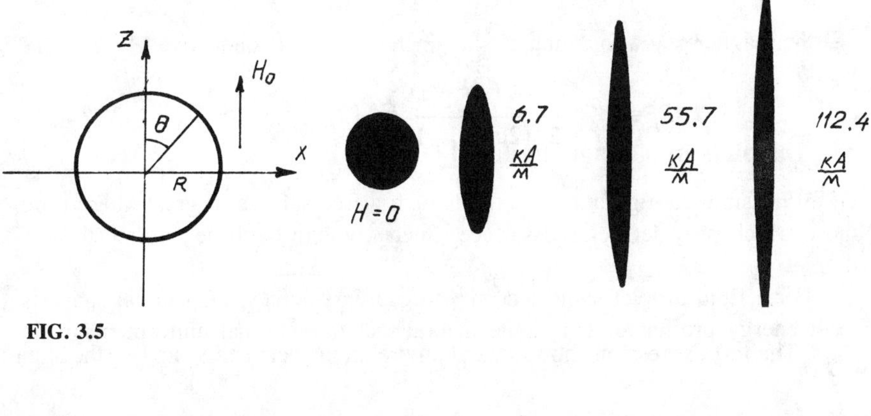

FIG. 3.5

and lower points of the sphere $z = \pm R$, as shown in Fig. 3.5, where the magnetic field $\vec{H}$ is normal to the surface. This differential pressure may only be balanced by increasing the surface curvature at the upper and lower points and by decreasing it at its side points. As a result, the droplet tends to extend along the field.

In a reverse situation, when a nonmagnetic fluid droplet is in a magnetic medium, the same mechanism should have compressed the droplet in a vertical direction. In this case, however, it is opposed by an external pressure gradient directed on the droplet surface to the side points and caused in the MF by field distortions due to the presence of the droplet. Note that in both cases the field inside the droplet is uniform. The joint action of both mechanisms is easy to calculate using the exact solution of the Maxwell equations for this case. For the linear law of magnetization $\vec{M} = \chi\vec{H}$

$$H_{er} = (H_0 - 2A/r^3)\cos\theta;\ H_{e\theta} = -(H_0 + A/r^3)\sin\theta;$$

$$H_{ir} = B\cos\theta,\ H_{i\theta} = -B\sin\theta$$

where

$$A = \frac{\mu_e - \mu_i}{2\mu_e + \mu_i} H_0 R^3,\ B = \frac{3\mu_e}{2\mu_e + \mu_i} H_0,\ \mu = \mu_0(1+\chi)$$

Using the ferrohydrostatic equation $\nabla P = \mu_0 M \nabla H$, obtain in fluid volumes

$$P_i = \text{const}_i;$$

$$P_e = \text{const}_e + \frac{\mu_0\chi_e}{2}\left(H_0 + \frac{A}{r^3}\right)^2 - \frac{3\mu_0\chi_e A}{2r^3}\left(2H_0 - \frac{A}{r^3}\right)\cos^2\theta$$

and the condition for the pressure jump on the sphere surface

$$P_i - P_e = 2\alpha/R - (1/2)\mu_0(M_{ir}^2 - M_{er}^2)$$

gives at $r = R$

$$P_i = D - \frac{9}{2}\frac{\mu_0^2\mu_e}{(2\mu_e + \mu_i)^2} H_0^2(\chi_i - \chi_e)^2\cos^2\theta$$

where

$$D = \text{const}_e + \alpha\frac{2}{R} + \frac{\mu_0\chi_e}{2}\frac{H_0^2}{(2\mu_e + \mu_i)^2}$$

The last expression shows that at any relations between χ_i and χ_e the situation is equivalent to a droplet with lower inside pressure at the upper and lower points of the surface, which, as already shown, extends the droplet along the field lines. It is worth noting that differential pressures at the upper (lower) and side points on the surface

$$\Delta P_i = \frac{9}{2} \frac{\mu_0^2 \mu_e}{(2\mu_2 + \mu_i)^2} H_0^2 (\chi_i - \chi_e)^2$$

are greater for a nonmagnetic droplet in MF than for a droplet of the same MF in a nonmagnetic fluid. They are related as $\mu_r(2 + \mu_r)^2/(2\mu_r + 1)^2$ (where $\mu_r = \mu/\mu_0$), while at large μ_r as $\mu_r/4$, i.e., the former is deformed more intensively at first than the latter.

Thus, the distortion of a uniform magnetic field in MF by the introduced nonmagnetic spherical droplets turns out to be a more essential reason for a change in the droplet shape than are the pressure jumps at the interface.

In Figs. 3.5 and 3.6, the experimental results show how the MF droplet extends along the magnetic field. In experiments, the magnetic fluid droplet was subjected to hydroimponderability in an aqueous glycerin solution. Criterial interpretation of the results allows the experiment to be plotted by one curve of the relative droplet length l/d_0 vs the dimensionless parameter $S = \mu_0 M^2 d_0/\alpha$. The experimental points correspond to various initial droplet diameters d_0 and magnetization M values for a prescribed value of the magnetic field intensity.

At small field intensities, the droplet shape is close to an ellipsoid. A rapid change of relative droplet sizes occurs at field intensities up to 100 kA/m. In the region $H > 100$ kA/m, the curves show saturation. In Fig. 3.6, this region corresponds to $s > 100$. The dependence depicted in this Figure is similar to $s > 100$ with the saturation onset in the fields of the order of 100 kA/m (cf. Fig. 2.2).

MF Droplet in a Nonuniform Field

In a nonuniform field, the droplet shape will in a great measure be determined by a magnetostatic force. In order to study this problem let us consider a situa-

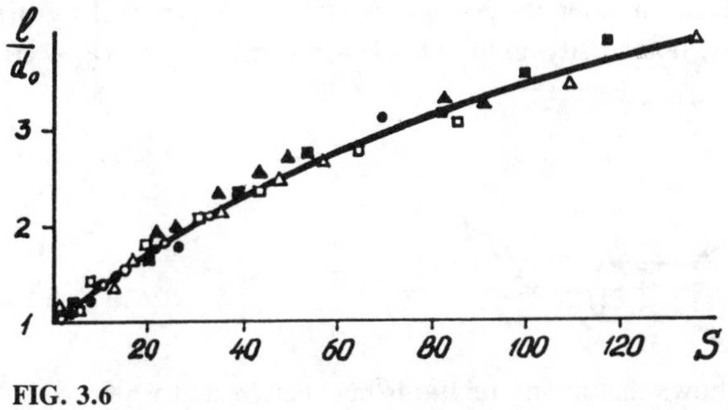

FIG. 3.6

tion when the presence of a droplet does not distort the external magnetic field and, besides, there are no effects due to a magnetic pressure jump at interfaces. These conditions are fulfilled and experimentally realized for a MF droplet around a live cylindrical conductor (Fig. 3.7). The free surface of the droplet $r = \xi(z)$ has an axial symmetry, and, therefore, does not distort the magnetic field of the current which has the form $H = [0, H_\varphi = I/(2\pi r), 0]$ in the entire space. Using the explicit form of the surface curvature in cylindrical coordinates, we shall arrive at the following equation to determine the droplet shape

$$(1+\xi'^2)^{-3/2}\left(\frac{1+\xi'^2}{\xi} - \xi''\right) = \frac{\mu_0}{\alpha}\int_0^{H(\xi)} M\,dH + C \qquad (3.27)$$

where the primes mean differentiation over z.

Taking account of the problem's symmetry, we can only consider the upper right-hand quarter of the droplet and write the boundary conditions for ξ as follows:

$$\xi(0) = \alpha,\ \xi'(0) = 0,\ \xi(z_0) = R,\ \xi(z_0) = -\mathrm{tg}\gamma \qquad (3.28)$$

The external maximum radius a and the half-length z_0 of the droplet are unknown and have to be estimated in the course of problem solution. The wetting angle γ is considered to be prescribed and the five unknown constants available (C, a, z_0 and two constants obtained at integrating equation (3.27)) are found from four boundary conditions (3.28) and the condition of a constant droplet volume

$$V/2\pi = \int_0^{z_0} (\xi^2 - R^2)\,dz = \mathrm{const} \qquad (3.29)$$

If the droplet is very thin, i.e., $\delta = (a - R)/R \ll 1$, then Eq. (3.27) acquires the most simple dimensionless form, correct to the terms of the order of

$$\partial^2\varepsilon/\partial y^2 - (\mathrm{Bo}_m - 1)\varepsilon = -C_1 \qquad (3.30)$$

in which the magnetic Bond number $\mathrm{Bo}_m = \mu_0 M G R^2/\alpha$ is determined by the characteristic magnetic field intensity gradient: $G = I/(2\pi R^2)$; $\varepsilon = (\xi - a)/(a - R)$; $y = z/R$.

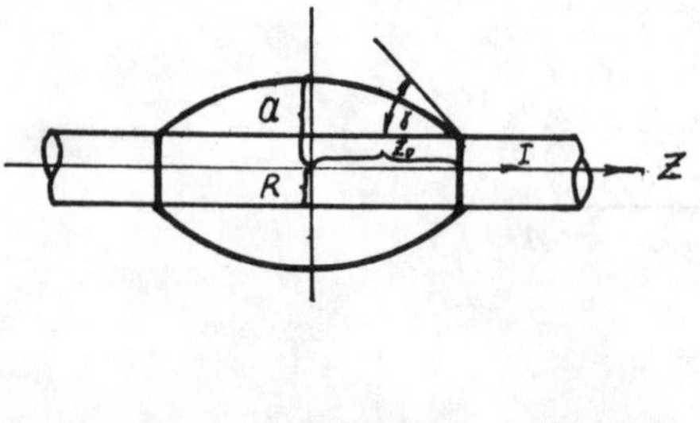

FIG. 3.7

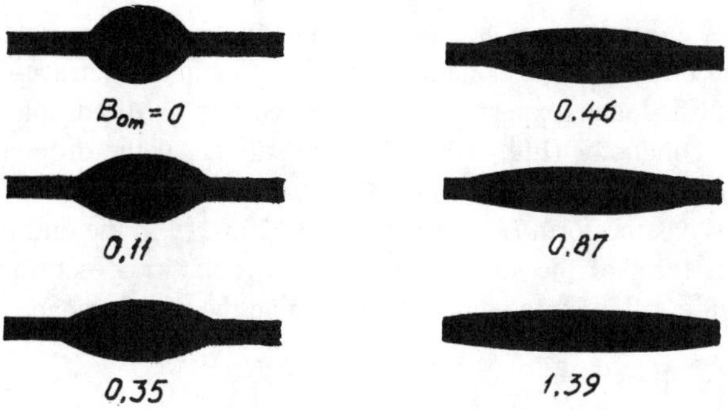

FIG. 3.8

This equation is easy to solve and at $\mathrm{Bo}_m > 1$, for example, yields

$$\left.\begin{aligned} &\varepsilon = \frac{\mathrm{ch}\beta y_0 - \mathrm{ch}\beta y}{\mathrm{ch}\beta y_0 - 1}\ ;\ \mathrm{th}(\beta y_0) = \left(1 + \frac{\beta^2 V}{4\pi R^3 \mathrm{tg}\,\gamma}\right)^{-1} \beta y_0\ ; \\ &\delta = \frac{\mathrm{tg}\gamma}{\beta}\,\frac{\mathrm{ch}\,\beta y_0 - 1}{\mathrm{ch}\,\beta y_0}\ ;\ C_1 = \frac{\beta^2 \mathrm{ch}\,\beta y_0}{\mathrm{ch}\beta y_0 - 1}\ ;\ \beta = \sqrt{\mathrm{Bo}_m - 1} \end{aligned}\right\}^{*} \tag{3.31}$$

At large values of the magnetic Bond number, we obtain from (3.31)

$$y_0 = z_0/R = \sqrt{\mathrm{Bo}_m}\, V/(4\pi R^3 \mathrm{tg}\gamma),\ \delta = \mathrm{tg}\gamma/\sqrt{\mathrm{Bo}_m}$$

In these relations $y_0 = z_0/R$ specifies the droplet half-length and δ, its maximum external radius a. The droplet length is seen to be in direct proportion to the square root of Bo_m and the maximum radius is inversely proportional to it. As Bo_m increases, the droplet contracts in a transverse direction and extends in a longitudinal direction. Other conditions being equal, the droplet radius is the greater the larger the wetting angle γ. This is illustrated by the plots in Fig. 3.8.

* See nomenclature on p. 211 for explanation of notation for trigonometric functions.

CHAPTER

FOUR

THERMOCONVECTIVE PHENOMENA

As already stated, nonisothermal flows of magnetic fluids as well as those with free surface, i.e., film flows, are of the greatest interest in the MF dynamics. In the former case, the thermomagnetic convection mechanism is most striking. In the latter, the magnetic field is responsible for the fluid dynamics by affecting the form and stability of its free surface. Both modes of flow are widely used in heat and mass exchangers and are of great practical interest.

4.1 THERMOMAGNETIC CONVECTION

The physics of thermomagnetic convection was discussed in 3.1 herein. It implies that, in a nonisothermal MF, cold layers of large magnetization are sucked into regions with a magnetic field of higher intensity, i.e., they move along the magnetic field intensity gradient displacing warmer (more heated) layers. The convection intensity is mainly determined by the modulus of the field intensity gradient and the temperature coefficient of magnetization. The higher their values, the more enhanced the convection is.

In zero approximation, referred to as noninductive one, the magnetic field changes caused by a nonisothermal fluid may be ignored and the field may be assumed to be prescribed. If we write γ for the fluid temperature gradient, then the magnetization gradient it causes will be $|\nabla M| = |dM/dT|\gamma = K\gamma$. Of the same order will be the field intensity gradient due to nonisothermal state of the fluid. When this gradient is much less than the external field gradient G, it may be ignored. The main condition for the nonconductive approximation to be applied is $G \gg K\gamma$. At $\gamma \sim 10^2$ K/m the value of $K\gamma$ is in the order of 10^4 A/m^2.

Mathematically, noninductive approximation means isolating the Maxwell equations from the complete set of equations for magnetic fluid thermomechanics. It is precisely in this approximation that thermomagnetic convection has been studied to-date.

In order to write down the equations for thermomagnetic convection, the equations of state must be substituted into the equations of motion and exclude the hydrostatic pressure gradient from our consideration. We obtain from equations (2.34)

$$\begin{aligned} &\rho^*[\partial \vec{v}/\partial t + (\vec{v}\nabla)\vec{v}] = -\nabla P' + \eta\Delta\vec{v} - \rho^*\beta_\rho\vartheta\vec{g} - \mu_0 K\vartheta\nabla H; \\ &\operatorname{div}\vec{v} = 0; \\ &\partial\vartheta/\partial t + \vec{v}\nabla\vartheta = \kappa\Delta\vartheta \end{aligned} \quad (4.1)$$

where $\vartheta = T - T^*$, $P' = P - P_0$, $\nabla P_0 = \rho^*\vec{g} + \mu_0 M^*\nabla H + \chi_r(H - H^*)\nabla H$, κ is thermal diffusivity. The equivalence of Archimedes' expulsive force $\rho^*\beta_\rho\vec{g}\vartheta$ and thermomagnetic force $\mu_0 K\nabla H\vartheta$ is most attractive. The magnetization temperature coefficient K is of the same significance as the fluid thermal expansion coefficient $\rho^*\beta_\rho$, while the magnetic field gradient ∇H is equivalent to the gravitational acceleration $\vec{g}$ but unlike the latter it may have various spatial configurations. This enriches thermomagnetic convection and makes it more controllable.

The introduction of scale quantities gives equations for thermomagnetic convection in the form more convenient to be analyzed and solved:

$$\left.\begin{aligned} &\partial\vec{v}/\partial t + (\vec{v}\nabla)\vec{v} = -\nabla P' + \Delta\vec{v} - (\mathrm{Gr}\,\vec{e}_g + \mathrm{Gr}_m\,\tilde{G}\vec{e}_G)\vartheta\ ; \\ &\operatorname{div}\vec{v} = 0\ ; \\ &\partial\vartheta/\partial t + \vec{v}\nabla\vartheta = \mathrm{Pr}^{-1}\Delta\vartheta \end{aligned}\right\} \quad (4.2)$$

The same notation for dimensionless and dimensional quantities should not cause any misunderstanding. A characteristic dimension is set as scales as in the thermal convection theory, for instance l:l^2/ν for time; ν/l for velocity; $\rho^*\nu^2/l^2$ for pressure; γl for temperature; γ the characteristic temperature gradient; ν the kinematic viscosity; $\vec{e}_g = \vec{g}/g$ and $\vec{e}_G = \nabla H|\nabla H$ are unit vectors in the direction of the gravitational force and thermomagnetic force (the latter may be the function of coordinates and time); $\tilde{G}$ is the dimensionless quantity of the magnetic field intensity gradient related to some characteristic value of the gradient $\tilde{G}^*$:$\tilde{G} = |\nabla H|/G^*$ and, in the general case, being the function of coordinates and time.

As is known, thermal gravitational convection is described by the dimensionless Grashof $\mathrm{Gr} = \beta_\rho g\gamma l^4/\nu^2$ and Prandtl $\mathrm{Pr} = \nu/k$ numbers. Thermomagnetic convection is described by dimensionless criteria referred to, by analogy, as the magnetic Grashof number, $\mathrm{Gr}_m = \mu_0 KG^*\gamma l^4/(\rho^*\nu^2)$. The ratio between the magnetic and gravitational Grashof numbers $Gr_m/Gr = \mu_0 KG/(\rho^* \beta_\rho g)$ yields a comparative estimate for the rate of magnetic convection. In magnetic fields with the intensity gradient of the order of 10^6 A/m^2, this ratio is of the order of 10, which evidences a possible appreciable predominance of thermomagnetic convection over gravitational convection.

The thermomagnetic convection equations are most vividly analyzed when the intensity gradient $|\nabla H| = G^*$ is constant and its direction, for example, coincides with that of the gravity force. Then the volume force in the equation of motion (4.2) assumes the form $Gr^+ \vartheta \vec{e}_g$ similar to the Archimedes' expulsive force responsible for thermal gravitational convection. In this case, the Grashof number is the sum of gravitational and magnetic numbers: $Gr^+ = Gr + Gr_m = (\beta_\rho g + \mu_0 KG^*/\rho^*)\gamma l^4/\nu^2$. In this situation, the gravitational and thermomagnetic convection mechanisms enhance each other. With the gravitational force and the magnetic field intensity gradient opposing each other, the above convection mechanisms act in opposite directions, thereby attenuating each other: $Gr^+ = Gr - Gr_m$.

This analysis may be regarded as gravitational analogy of thermomagnetic convection. This analogy allows all the relations obtained for thermal gravitational convection to be applied to thermomagnetic convection. Thus, the free convection heat transfer equations of the type $Nu = f(Gr)$ still hold, with the substitution of the total Grashof number $Nu = f(Gr^+)$. Having in mind that the values of the magnetic Grashof number may be by an order of magnitude higher than the gravitational number, it is evident that convective heat transfer may be enhanced by the thermomagnetic mechanism. Besides, under the conditions of weightlessness the thermomagnetic mechanism remains the only one to ensure free-convection heat transfer.

A complex MF convective motion and rather intricate convective heat transfer regularities may be observed in magnetic fields of complex spatial configuration. Vast material on this subject is contained in monographs [7, 9]. In many cases, however, the gravitational analogy turns out to be useful for qualitative analysis of the results.

All the above-said supports the possibility of local enhancement of heat transfer in a fluid, with strong local magnetic field nonuniformities set up in it. It is also possible to model, by means of magnetic fields, convective motion in the systems affected by different kinds of acceleration and complex gravitational fields. The latter situations include, for example, convection in self-gravitating masses.

As a particular example consider a thermoconvective flow of a weightless magnetic fluid in an annular channel formed by two coaxial cylinders, with current I in the internal channel inducing a radially nonuniform magnetic field $H = H\varphi = I/(2\pi r)$, $\nabla H = -(I/2\pi r^2)\vec{e}_r$ (Fig. 4.1). A longitudinal temperature gradient is kept constant at the channel boundaries $r = R_1; R_2$. It should be noted that this problem allows an accurate one—dimensional steady-state solution to the complete set of MF thermomechanics equations, including the Max-

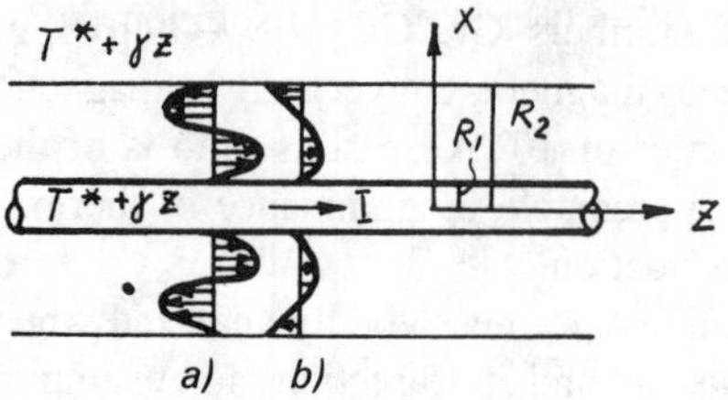

FIG. 4.1

well equation, with all the boundary conditions for a magnetic field to be satisfied.

The solution to the problem is sought in the form $\vec{v} = [0, 0, v(x)]$; $\vartheta = z + \theta(x)$, $\vec{H} = [0, H(x)0]$, $\vec{M} = [0, M(x,z), 0]$ in dimensionless coordinates z and $x = (r - R_1)/(R_2 - R_1)$. The channel width $l = R_2 - R_1 = \nabla R$ is the length scale; $\gamma\Delta R$ is the temperature scale; $G^* = I/(2\pi R_1^2)$ is the characteristic field gradient. In the instant under consideration, $\vec{e}_G = -\vec{e}_x = -\vec{e}_r$, $\tilde{G} = 1/(1 + \delta\chi)^2$ where $\delta = (R_2 - R_1)/R_1$ is the relative layer thickness.

Equations (4.2) for this problem yield

$$\frac{\partial P'}{\partial x} = \frac{\mathrm{Gr}_m}{(1+\delta x)^2}\vartheta; \quad \frac{\partial P'}{\partial z} = \frac{1}{1+\delta x}\frac{\partial}{\partial x}\left[(1+\delta x)\frac{\partial v}{\partial x}\right];$$

$$v = \mathrm{Pr}^{-1}\frac{1}{1+\delta x}\frac{\partial}{\partial x}\left[(1+\delta x)\frac{\partial \theta}{\partial x}\right]; \quad \mathrm{Gr}_m = \frac{\mu_0 K I \gamma \Delta R^4}{2\pi R_1^2 \rho^* \nu^2}$$

Eliminating the pressure from the first two equations by cross-differentiation gives the equation for velocity

$$\frac{\partial}{\partial x}\left\{(1+\delta x)^{-1}\frac{\partial}{\partial x}\left[(1+\delta x)\frac{\partial v}{\partial x}\right]\right\} = \frac{\mathrm{Gr}_m}{(1+\delta x)^2}$$

with its general solution

$$v = -\frac{\mathrm{Gr}_m}{\delta^3}\left[(1+\delta x) + C_1(1+\delta x)^2 + C_2\ln(1+\delta x) + C_3\right] \tag{4.3}$$

Integration of the equation for temperature with the velocity profile known, presents no difficulties and yields another two arbitrary constants to be estimated from the boundary conditions for temperature: $\theta = 0$ at $x = 0; 1$.

The constants C_1, C_2 and C_3 are found from two boundary conditions for velocity and the fluid flow rate prescribed through the channel cross-section. If the channel is closed, the flow rate is zero

$$\int_{R_1}^{R_2} rv\,dr = 0$$

With this in view, for solid channel walls ($v = 0$ at $x = 0; 1$) solution (4.3) becomes

$$\left.\begin{aligned} v = &-\frac{\mathrm{Gr}_m}{\delta(2+\delta)}\Big\{x - x^2 + C[x(2+\delta x)\ln(1+\delta) - \\ &-(2+\delta)\ln(1+\delta x)]\Big\}; \\ C = &(\delta/3)\,[(2+\delta+\delta^2)\ln(1+\delta) - \delta(2+\delta)]^{-1} \end{aligned}\right\} \qquad (4.4)$$

The solution obtained describes a closed convective flow along the temperature gradient at the walls of the internal cylinder and opposite in direction to the external one, as shown in Fig. 4.1a. The velocity modulus of convective flow is in direct proportion to the magnetic Grashof number.

When the channel width is small ($\delta \ll 1$), (4.4) becomes the known solution for a plane-parallel free-convective flow in a horizontal layer with longitudinal temperature gradient under the action of the gravitational force.

$$v = \frac{\mathrm{Gr}_m}{12}(2x^3 - 3x^2 + x),$$

where Gr_m is an ordinary Grashof number. This is the case of gravitational analogy.

Thermocapillary Convection

Solution (4.3) also provides for a situation when the external fluid surface is free. As found above, the free cylindrical MF surface in the field of a line conductor is quite natural. Then the boundary condition for velocity on this surface, with allowance for the surface tension coefficient α vs. temperature ($\alpha = \alpha^* - \beta_\alpha(T - T^*)$) from (2.48), will have the form

$$\eta(\partial v/\partial r)_{r=R_2} = (\partial\alpha/\partial T)(\partial T/\partial z) = -\beta_\alpha\gamma \qquad (4.5)$$

The free-surface tension coefficient gradient specifies the thermocapillary convection mechanism described by the Marangoni number $\mathrm{Ma} = \beta_\alpha\gamma l^2(\eta k)$. If the thermocapillary convection mechanism is predominant ($\mathrm{Ma} \gg \mathrm{Gr}_m$), then from (4.3) follows velocity distribution in the channel (cf. Fig. 4.1b)

$$v = \{x(2+\delta x) + D[2(1+\delta)^2\ln(1+\delta x) - \delta x(2+\delta x)]\} \cdot \{-\mathrm{Ma}/[2(1+\delta)]\}$$

where

$$D = -\delta(2+\delta)^2[4(1+\delta)^4\ln(1+\delta) - \delta(2+\delta)(2+6\delta+3\delta^2)]^{-1} \qquad (4.6)$$

In this case, the velocity profile has only one extremum, the convection rate being specified by the Marangoni number. In such a situation, the magnetic field ensures the existence of a fluid volume with a cylindrical free surface.

The thermocapillary force as applied to MFs is of research interest. This is because of a variety of equilibrium fluid surface shapes and the effects that appear due to the joint action of thermocapillary and magnetostatic forces which will be discussed below.

It should be noted in conclusion that so far we have dealt with the magnetic field intensity gradient caused by nonuniformity of the external field. Certainly, the heaviest gradients may thus be achieved. However, even in an external uniform magnetic field inside the fluid there may be field gradients due, firstly, to the bending of field lines at the vessel walls confining the fluid and, secondly, to the nonisothermal state of the fluid and, hence, nonuniformity of its magnetization. In these instances, the field in the fluid is found by solving the Maxwell equation. As temperature distribution is, to a great extent, determined by fluid motion, all the equations of ferrohydrodynamics become interrelated.

While the direction of the magnetic field intensity vector is unimportant in the determination of the volume magnetic force $\mu_0 M \nabla H$, it is of great significance in Maxwell equations. In order to relate the magnetic field intensity and fluid magnetization intensities, the equation div $\vec{B} = 0$ is used. Bearing in mind that $\vec{M} = [M(H,T)/H]\vec{H}$, we may represent it as

$$\left(1 + \frac{M}{H}\right) \operatorname{div} \vec{H} = \left(\frac{M}{H} - \frac{\partial M}{\partial H}\right) \vec{e}_H \cdot \nabla H - \left(\frac{\partial M}{\partial T}\right) \vec{e}_H \cdot \nabla T \tag{4.7}$$

where $\vec{e}_H = H/H$ is the unit vector in the direction of the field. Together with rot $\vec{H} = 0$, equation (4.7) must close the set of equations (4.1) in the new statement.

In a majority of cases, the exact solution of (4.7) is impossible. It is therefore advisable that a simplified solution be sought. One such solution may be when the magnetic field has a distinct constant intensity component H^* whose value considerably exceeds the field intensity change ΔH in the region under consideration $H^* \gg \Delta H \sim Gl$. The same is true of the fluid magnetization $M^* \gg \Delta M$. This approximation may be referred to as low magnetic field nonuniformity approximation. The coefficients in Eq. (4.7) may then be assumed constant ($M/H \approx M^*/H^* = \chi_s$, $\partial M/\partial H = \chi_r$, $\partial M/\partial T = -K$) and written down in the form

$$\operatorname{div} \vec{H} = \frac{\chi_s - \chi_r}{1 + \chi_s} \vec{e}_H^{\,*} \nabla H + \frac{K}{1 + \chi_s} \vec{e}_H^{\,*} \cdot \nabla T \tag{4.8}$$

where the vector $\vec{e}_H^{\,*} = \vec{H}^*/H^*$ may also be considered a constant unit vector in the direction of the constant field component.

By making this equation dimensionless just as system (4.1) and setting the quantity G^*l as a field intensity scale, represent it as

$$A\text{div}\vec{H} - (A - 1)\vec{e}_H^{\,*}\nabla H - A_1\vec{e}_H^{\,*} \cdot \nabla\vartheta = 0 \tag{4.9}$$

By introducing the field potential $\vec{H} = \nabla\Phi$, we may automatically satisfy $\text{rot}\vec{H} = 0$.

In (4.9), there have appeared two new dimensionless parameters:

$$A = (1 + \chi_s)/(1 + \chi_r),\ A_1 = K_\gamma/[(1 + \chi_r)G^*]$$

The first parameter specifies the degree of nonlinearity of the fluid magnetization law. With the linear law $\vec{M} = \chi\vec{H}$, we have $\chi_r = \chi_s = \chi$, $A = 1$ and the second term in Eq. (4.9) disappears. The parameter A may assume the value close to unity in a state of fluid saturation at $H \gg M$. Then χ_ρ; $\chi_s \ll 1$ and $A \approx 1$. The second parameter, A, characterizes the relative contribution of non-isothermal fluid to the total field intensity gradient. At $A_1 \ll 1$, the last term in Eq. (4.9) may be ignored. Thus, temperature is eliminated from the Maxwell equations and they are isolated from the other equations. So, the condition $A_1 \ll 1$, i.e., $G^* \gg (1 + \chi_\rho)K\gamma$, assesses the applicability of noninductive approximation, as was shown above.

For a wide range of problems, Eq. (4.9) allows the temperature effect on magnetic fluid distribution in the fluid to be taken account of in a first approximation.

4.2 THERMOCONVECTIVE INSTABILITY

As was elucidated in 3.1, mechanical equilibrium may take place in a nonisothermal MF in case of certain directions of the magnetic field and temperature gradients. However, the effect of thermogravitational and thermomagnetic mechanisms may lead to unstable equilibrium under certain conditions resulting in a convective motion of the fluid. This phenomenon is called thermoconvective instability. In MFs, thermoconvective instability acquires specific properties owing to thermomagnetic convection mechanism. Thermoconvective instability has a threshold behaviour, i.e., it appears at certain critical temperature and field gradients. Its onset is determined by critical Grashof Gr and Rayleigh Ra numbers $\text{Ra} = \text{Gr} \cdot \text{Pr}$ dependent on the channel geometry and equilibrium temperature, gravitational and magnetic field configuration as well as on the nature of perturbations in the system.

Most often equilibrium stability is analysed with respect to infinitesimal perturbations that allow linearization of the input equations. Let, in a state of equilibrium, the quantities characterizing the situation be $\vec{v}_0 = 0$, P_0, ρ_0, T_0, $\vec{H}_0$, $\vec{M}_0$. The perturbations of velocity $\vec{v}$, pressure P', density ρ', temperature T', magnetic field H' and $\vec{h}$, magnetization M' and m are superimposed on this equilibrium. In an excited (perturbed) state $P = P_0 + P'$, $T = T_0 + T'$, $\rho = \rho_0 + \rho'$, $H = H_0 + H'$, $\vec{H} = \vec{H}_0 + \vec{h}$, $M =$

$M_0 + M'$, $\vec{M} = \vec{M}_0 + \vec{m}$. In an approximation linear relative to perturbations, the set of equations (2.34) becomes

$$\left.\begin{aligned}
&\rho\,\partial\vec{v}/\partial t = -\nabla\tilde{P}' + \eta\Delta\vec{v} - \beta_\rho\vec{g}T' - \mu_0 KT'\nabla H_0 + \mu_0 KH'\nabla T_0\,;\\
&\tilde{P}' = P' + \mu_0 MH';\\
&\operatorname{div}\vec{v} = 0;\ \partial T'/\partial t + \vec{v}\nabla T_0 = \kappa\Delta T', \quad \operatorname{rot}\vec{h} = 0\,;\\
&\operatorname{div}\left[\left(1 + \frac{M_0}{H_0}\right)\vec{h} + \left(\frac{\partial M}{\partial H} - \frac{M_0}{H_0}\right)\frac{\vec{H}_0}{H_0}H' - K\frac{\vec{H}_0}{H_0}T'\right] = 0\,;\\
&H' = (\vec{H}_0\vec{h})\,/\,H_0,\ M' = -KT' + (\partial M/\partial H)\,H';\\
&\vec{m} = \frac{M_0}{H_0}\vec{h} + \left(\frac{\partial M}{\partial H} - \frac{M_0}{H_0}\right)\frac{\vec{H}_0}{H_0}H' - K\frac{\vec{H}_0}{H_0}T'
\end{aligned}\right\} \quad (4.10)$$

In the case under consideration, the perturbations in all the quantities are due to convective motion arising in the field. In particular, perturbations in the magnetic field intensity H' are caused by temperature perturbations T'. While the order of the latter is determined by the temperature gradient γl, magnetic field perturbations are characterized by the quantity $K\gamma l/(1 + \chi_r)$ which is assumed to be a scale for them. The scales of other quantities remain as previously. By taking the Boussinesq and low magnetic field nonuniformity approximations and introducing the dimensionless field perturbation potential $\vec{h} = (K\gamma l/\mu_r)\psi$, the system of equations (4.10) may be written in the form

$$\left.\begin{aligned}
&\partial\vec{v}/\partial t = -\nabla\tilde{P}' + \Delta\vec{v} - \mathrm{Gr}\,\vartheta\vec{e}_g - \mathrm{Gr}_m\tilde{G}\vartheta\vec{e}_G + DH'\tilde{\gamma}\vec{e}_\gamma\,;\\
&\operatorname{div}\vec{v} = 0;\ \partial\vartheta/\partial t + \tilde{\gamma}\,\vec{v}\vec{e}_\gamma = \mathrm{Pr}^{-1}\Delta\vartheta\,;\\
&A\Delta\psi + (1 - A)\,\vec{e}_H^{\,*}\,\nabla H - \vec{e}_H^{\,*}\nabla\vartheta = 0\,;\ H' = \vec{e}^{\,*}\nabla\psi\,,
\end{aligned}\right\} \quad (4.11)$$

where

$$\vec{e}_g = \vec{g}/g\,,\ \vec{e}_G = \nabla H/|H_0|\,;\ \vec{e}_\gamma = \nabla T_0/|\nabla T_0|\,,\ \vec{e}_H^{\,*} = \vec{H}^*/H^*;$$
$$\tilde{G} = |\nabla H_0|/G^*,\quad \tilde{\gamma} = |\nabla T_0|/\gamma^*$$

The dimensionless parameter $D = \mu_0 K^2\gamma^2 l^4/[\rho^* \nu^2(1 + \chi_r)$ that appeared in the equation of motion explicitly describes the contribution of magnetic field perturbations H', caused by the motion of a nonisothermal fluid, to the magnetic force. It is equivalent to the magnetic Grashof number Gr_m with the characteristic field gradient $K\gamma/\mu_r$. The condition $\mathrm{Gr}_m \gg D$ coincides with the condition of noninductive approximation $G \gg K/\mu_r$ and means that the field perturbations due to the fluid motion may be neglected. At the same time, without equilibrium field gradient ($\tilde{G} = 0$), the last term in the equation of motion becomes predominant and describes the effect of uniform magnetic fields on thermoconvective processes in magnetic fluids. Numerical estimates show that the parameter D may assume values between 10^3 and 10^4. These values correspond, for example, to critical Grashof numbers, which points to a possible essential effect of uniform magnetic fields on convective stability of the fluid.

If the equilibrium temperature and magnetic field gradients are constant and parallel to gravitational acceleration, i.e., they are opposite in direction to the vertical axis z ($e_g = e_G = e\gamma = -1$, $\tilde{G} = 1$, $\tilde{\gamma} = 1$), Eqs. (4.11) may be written retaining only one vertical velocity component v_z and excluding the pressure

$$\left.\begin{aligned} &\partial \Delta v_z/\partial t = \Delta\Delta v_z + (\mathrm{Gr} + \mathrm{Gr}_m)\Delta_1 \vartheta - D(\vec{e}^{\,*}_H \cdot \nabla)\Delta_1 \psi ;\\ &\partial \vartheta/\partial t = \mathrm{Pr}^{-1}\Delta\vartheta + v_z ;\\ &A\Delta\psi + (1-A)(\vec{e}^{\,*}_H \cdot \nabla)^2\psi - \vec{e}^{\,*}_H \nabla\vartheta = 0;\\ &\Delta_1 = \partial^2/\partial x^2 + \partial^2/\partial y^2 \end{aligned}\right\} \quad (4.12)$$

In the case of noninductive approximation ($D = 0$), we come across gravitational analogy, and the critical values of the Grashof number ($\mathrm{Gr}^+ = \mathrm{Gr} + \mathrm{Gr}_m$) are the corresponding critical Grashof numbers for normal thermogravitational instability. Thus, for a horizontal plane-parallel layer $\mathrm{Ra}^+_* = \mathrm{Gr}^+_* \cdot \mathrm{Pr} = 1708$ (asterisk * is a subscript for critical values). For the considered gradient direction, the thermomagnetic mechanism decreases fluid stability, i.e., reduces the critical values of the temperature gradient. For a horizontal layer we have

$$\gamma_* = 1708 \frac{\kappa \nu}{(\beta_\rho g + \mu_0 K G^*/\rho^*) l^4}$$

For an opposite direction of the field gradient ($\vec{e}_g = \vec{e} = -\vec{e}_G$), the thermogravitational and thermomagnetic mechanisms act in opposed directions, and fluid stability is increased. For this example

$$\gamma_* = 1708 \frac{\kappa \nu}{(\beta_\rho g - \mu_0 K G^*/\rho^*) l^4}$$

If $\mu_0 K G^*/\rho^* > \beta_\rho g$, the thermomagnetic mechanism completely suppresses the thermogravitational instability of the fluid.

In the absence of gravitational force ($g = 0$, $\mathrm{Gr} = 0$), the thermomagnetic mechanism is the only reason for convective instability that does take place if the directions of the field and temperature gradients coincide ($\vec{e}_G \uparrow\uparrow \vec{e}_\gamma$) (cf. 3.1). Otherwise ($\vec{e}_G \uparrow\downarrow \vec{e}_\gamma$), equilibrium of the fluid is stable.

Whereas gravitational analogy allows one to understand many laws of convective MF instability in gradient magnetic fields, the role played by an equilibrium magnetic field is far from being that clear. After the state of equilibrium is discontinued and convective motion sets in, the equilibrium temperature field in the fluid is distorted, thus leading to the distortion of the equilibrium uniform magnetic field. The problem concerns the effect of these field perturbations on the fluid stability. Temperature distortions of the magnetic field are described by the term $e_H^* \nabla \vartheta$ in the Maxwell equations of systems (4.11) and (4.12). This term is other than zero only when temperature perturbation gradients with the projection onto the field direction appear in the fluid. In other words, the field is only disturbed by temperature perturbations that are nonuniform along the field. The uniform magnetic field, in turn, affects only those perturbations whose nonuniformity coincides with its direction.

Generally speaking, this effect may be of two kinds: it may either enhance or damp temperature disturbances in the fluid. The most general considerations stemming from the entire electrodynamic and magnetohydrodynamic practice evidence that the stationary magnetic field resists the factors causing its change. It should be expected in the situation under study that the uniform magnetic field will damp the temperature perturbations causing its disturbance, i.e., it will increase the stability of the system and render it immune to perturbations that are nonuniform along the direction of the field. In this case, the magnetic field creates the anisotropy of the properties of the system under consideration relative to the direction of perturbations leading to its instability. The direction normal to the field is most detrimental. The field does not exert any effect on the perturbations in this direction.

Of great interest are the possible patterns of a convective fluid flow that develop owing to convective instability. Most preferred for consideration are the flows periodic only in the direction perpendicular to the field, i.e., rolls with their axes parallel to the field. It is well known that cellular convection (Benard cells) is realized in infinite fluid layers under no-field conditions.

The above involves the situations when perturbations most detrimental in the direction of their propagation are degenerated. Horizontal plane-parallel layers heated from below refer, for example, to such degenerated geometries. The uniform magnetic field in the layer plane eliminates this degeneration. When degeneration has already been eliminated, for example, by the layer boundaries or the kind of heating, the uniform magnetic field hinders the development of convective disturbances to be able to increase the system stability threshold. As an example, consider two particular situations.

Horizontal Free-Boundary Layer

A horizontal plane-parallel fluid layer with constant temperature difference at its boundaries is a most wide-spread geometry in the study of thermoconvective processes. Of particular methodological importance is the case of free layer boundaries. Although such a case is hard to realize experimentally, theoretically we are able to obtain critical relations rather simple to analyze. These relations include all specific qualitative features of the phenomenon under consideration. The geometry of the problem in question is shown in Fig. 4.2. The uniform magnetic field $\vec{H}^*$ inside the fluid coincides with the external uniform field since it is tangential to interfaces everywhere. The equilibrium temperature gradient in the fluid $\gamma = (T_1 - T_2)/l$ is regarded as positive when the layer is heated from below ($T_1 > T_2$).

The problem is described by the equations following from (4.12) at

$$\mathrm{Gr}_m = 0, \quad \vec{e}_H^{\,*} = [1, 0, 0]:$$

$$\partial(\Delta v_z)/\partial t = \Delta\Delta v_z + \mathrm{Gr}\Delta_1\vartheta - D\Delta_1 \partial\psi/\partial x;$$

$$\partial\vartheta/\partial t = \mathrm{Pr}^{-1}\Delta\vartheta + v_z;$$

$$\partial^2\psi/\partial x^2 + A(\partial^2/\partial y^2 + \partial^2/\partial z^2)\psi - \partial\vartheta/\partial x = 0$$

and by the boundary conditions at $z = 0; 1$

$$v_z = \partial^2 v_z/\partial z^2 = \vartheta = \psi = 0$$

The layer stability is considered relative to normal disturbances propagating in its plane, i.e., proportional to $\exp(-\omega t + i\vec{k}\vec{r})$, $\vec{k} = [k_x, k_y, 0]$. The problem has the solution of the form:

$$v_z = v_{za}\sin(n\pi z)\exp(-\omega t + i\vec{k}\vec{r}), \quad \vartheta = \vartheta_a \sin(n\pi z)\exp(-\omega t + i\vec{k}\vec{r}),$$
$$\psi = \psi_a \sin(n\pi z)\exp(-\omega t + i\vec{k}\vec{r}), \quad n = 1, 2, 3 \ldots$$

with a corresponding equation for the frequency ω providing the nontrivial solutions:

$$\omega = \frac{1+\mathrm{Pr}}{2\mathrm{Pr}}(n^2\pi^2 + \kappa^2) \pm \left\{\left(\frac{\mathrm{Pr}-1}{2\mathrm{Pr}}\right)^2 (n^2\pi^2 + \kappa^2)^2 + \right.$$
$$\left. + \frac{\mathrm{Ra}\,\kappa^2}{\mathrm{Pr}(n^2\pi^2 + \kappa^2)} - \frac{D_p \kappa_x^2 \kappa^2}{\mathrm{Pr}(n^2\pi^2 + \kappa^2)[\kappa_x^2 + A(\kappa_y^2 + n^2\pi^2)]}\right\}^{1/2} \qquad (4.13)$$

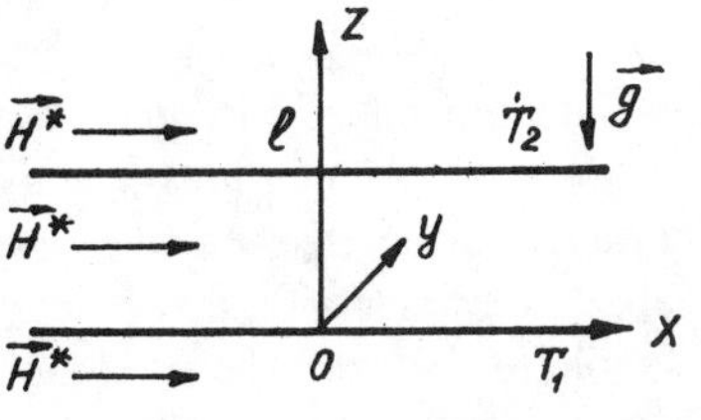

FIG. 4.2

where

$$\mathrm{Ra} = \mathrm{Gr} \cdot \mathrm{Pr}\ ,\quad D_p = D \cdot \mathrm{Pr}\ ,\quad \kappa^2 = \kappa_x^2 + \kappa_y^2$$

The relation so obtained points, above all, to the validity of the principle of changing stability in the situation discussed, i.e., all fluctuational disturbances with Im $\omega \neq 0$ have a positive decrement (Re $\omega > 0$) and attenuate in time.

The negative real values of the frequency, corresponding to the increase of disturbances with time, appear in (4.13) at

$$\mathrm{Ra} \geqslant \frac{(n^2\pi^2 + \kappa^2)^3}{\kappa^2} + \frac{D_p \kappa_x^2}{[\kappa_x^2 + A(\kappa_y^2 + n^2\pi^2)]} \qquad (4.14)$$

which is the condition of convective layer instability taking place only at heating the layer from below (Ra > 0). The second term in (4.14) is due to magnetic field effect and, being positive, increases the Rayleigh number for field instability. However, this is only true of the disturbances with x-projection of the wave number, i.e., nonuniform along the magnetic field direction. The magnetic field does not exert any effect on distributions with $k_x = 0$; area of their instability is the same as for a normal liquid with the boundary below than for the former ones. In other words, minimization of instability condition (4.14) over k_x, k_y and n gives the following critical values: $\mathrm{Ra}_* = (27/4)\pi^4$, $k_{x*} = 0$, $k_{y*} = \pi/\sqrt{2}$, $n_* = 1$. Hence, it follows in compliance with the above that, in the case under consideration, the uniform magnetic field does not change the instability onset threshold (as compared to normal fluid) but eliminates the available degeneration in the direction of most detrimental disturbances and forms convective motion as rolls with field-parallel axes instead of structures periodic in both directions in the plane of the layer (Benard cells).

Vertical Layer Heated From Below

The system, in which degeneration in the direction of most detrimental thermoconvective disturbances is eliminated by its very geometry, is a vertical plane-parallel fluid layer with a vertical temperature gradient maintained at its boundaries (Fig. 4.3). As the temperature in a state of mechanical equilibrium is not

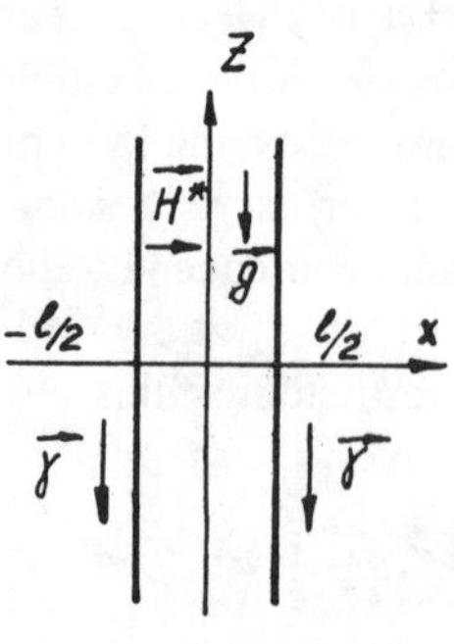

FIG. 4.3

constant, magnetization in the fluid is distributed vertically $M_{x0} = M^*(T^*, H^*) + K\gamma z$. Upon the instability onset, plane-parallel convective motion $v_z(x)$ is developing in the layer to shape temperature $\vartheta(x)$ and magnetic $\psi(x)$ fields. Thus, the primary uniform transverse magnetic field $\vec{H}^* = [H^*, 0, 0]$ appears to be inevitably disturbed. Solution to equations (4.12) for the problem with the boundary conditions at $x = \pm 1/2$

$$v_z = \vartheta = \partial\psi/\partial x = 0$$

is of the form

$$v_z = v_{za}\cos[(2n+1)\pi x]\exp(-\omega t);$$
$$\vartheta = \vartheta_a \cos[(2n+1)\pi x]\exp(-\omega t);$$
$$\psi = \psi_a \sin[(2n+1)\pi x]\exp(-\omega t),\ n = 0, 1, 2\ldots$$

and the range of eigen values ω is determined by the equation

$$\omega = \pi^2(2n+1)^2(1+\mathrm{Pr})/(2\mathrm{Pr}) \pm \sqrt{\pi^4(2n+1)^4(1-\mathrm{Pr})^2/4\mathrm{Pr}^2 + (\mathrm{Ra}-D_p)/\mathrm{Pr}} \qquad (4.15)$$

As the temperature gradient γ is the main physical quantity for the convective instability of the system, the further analysis will be based on this temperature gradient. Because γ enters both the Rayleigh number Ra and D_ρ let us introduce a new dimensionless parameter $N = D_p/\mathrm{Ra}^2 = \mu_0 v_k K^2/[\rho^*\mu_r(\beta_\rho g)^2 l^4]$ independent of γ. On account of the validity of the stability changing principle the neutral stability curve is determined by zero roots ω in (4.15):

$$\mathrm{Ra} = (2n+1)^4\pi^4 + \mathrm{Ra}^2 N$$

Critical Rayleigh numbers correspond to the lowest perturbation mode with $n = 0$ and are

$$\mathrm{Ra}_* = \pi^4 + \mathrm{Ra}_*^2 N \qquad (4.16)$$

Of interest here is the term $\mathrm{Ra}_*^2 N$ which appears due to the magnetic field effect and increases the critical Rayleigh numbers. Therefore, in the situation when the convective motion pattern is completely determined by the system geometry and inevitably disturbs the magnetic field, the latter exerts a stabilizing impact by increasing the critical Rayleigh numbers and, consequently, the critical temperature gradients.

However, the magnetic field effect is by no means restricted by this circumstance alone. Solving equation (4.16) relative to Ra_* gives

$$\mathrm{Ra}_* = 2\pi^4/(1 \pm \sqrt{1 - 4\pi^4 N})$$

whence it can be seen that the domain of real roots of equation (4.16), corresponding to a situation of instability, is restricted by the condition $N < 1/4\pi^4$. At $N > 1/4\pi^4$, equation (4.16) has no real roots, there are no complex frequencies with negative real part in the range of frequencies and all the disturbances attenuate. This may imply that the uniform magnetic field completely stabilizes the system under consideration at certain values of the parameters ($N > 1/4\pi^4$).

In Fig. 4.4, the region of instability is restricted by the curves $Ra_*^+ = 2\pi^4/(1 + \sqrt{1 - 4\pi^4 N})$ at the bottom and $Ra_*^- = 2\pi^4/(1 - \sqrt{1 - 4\pi^4 N})$ at the top. At $N = 1/4\pi^4$, these curves converge to form a point $Ra_* = 2\pi^4$ and, at $N > 1/4\pi^4$, the region of instability disappears. Thus, the uniform magnetic field increases the threshold values of the Rayleigh number from π^4 to $2\pi^4$ and also restricts the region of instability from right and above. The possibility of complete stabilization of the layer by the uniform magnetic field as well as the presence of the upper instability region boundary, are attributed to the increase of the stabilizing effect of the field with the growth of the temperature gradient in proportion to its square ($D_p \sim \gamma^2$), the convection moving force being proportional to its first power ($Ra \sim \gamma$).

The above-said allows the conclusion that the magnetic field is an effective means of controlling the convective stability of MFs.

4.3 INTERNAL MAGNETIC WAVES

The MF wave phenomena possess a number of specific features, among them the existence of new kinds of waves and the changing conditions of the propagation and generation of the known waves.

Internal magnetic waves represent a combined propagation of velocity, tem-

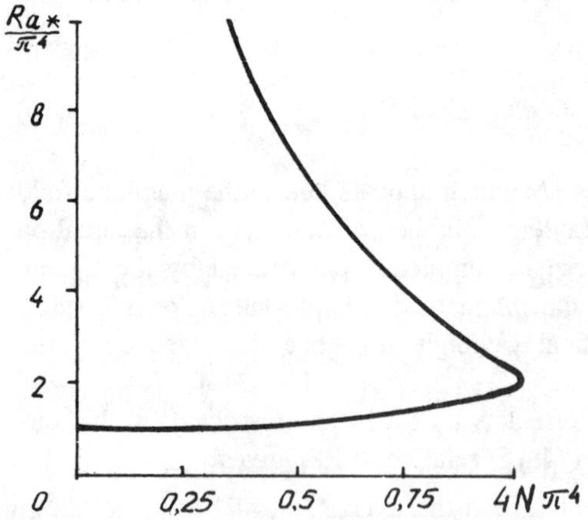

FIG. 4.4

perature, density, magnetic field intensity and continuum magnetization waves. They have an analog in the normal liquid, namely, internal gravitational waves [2]. In accordance with the type of the field responsible for the propagation mechanism, they are called internal magnetic waves.

As found in 3.1, a restoring force acts on a fluid volume element, shifted from equilibrium, in a nonisothermal magnetic fluid with opposite field and temperature gradients. Under the action of this force and inertial forces the volume element may vibrate close to equilibrium position, and its vibrations are transmitted in the fluid as internal magnetic waves.

Mathematical study of low-amplitude internal magnetic waves relies on a linearized set of equations (4.10). Their propagation is considered against the background of mechanical equilibrium in a nonisothermal fluid when the field $\vec{G}$ and temperature $\vec{\gamma}$ gradients are assumed to be constant and opposite ($\vec{G} \uparrow\downarrow \vec{\gamma}$). The gravitational force is supposed to be zero ($\vec{g} = 0$). In the case of an ideal fluid ($\eta = k = 0$) set of Eqs. (4.10) has a solution in the form of plane waves $\vec{v}, T', \tilde{p}', H', \vec{h}' \sim \exp i(\vec{k}\vec{r} - \omega t)$ described by the dispersion equation,

$$\omega_M^2 = \left(\frac{\mu_0 K G \gamma}{\rho} + \frac{\mu_0 K^2 \gamma^2}{\rho(1+\chi_r)} \cdot \frac{1}{1 + A\,\mathrm{tg}^2 \varphi_H}\right) \sin^2 \varphi_G \qquad (4.17)$$

where φ_H and φ_G are the angles between the wave vector $\vec{k}$ and, correspondingly, between the direction of its field $\vec{H}^*$ and gradient $\vec{G}$. The values of wave variables are related as

$$\vec{v}\vec{\gamma} = i\omega T', \; H' = KT'/[(1 + \chi_r)(1 + A\,\mathrm{tg}^2\, \varphi_H)]$$

These are internal magnetic waves.

From the continuity equation div $\vec{v} = 0$ it follows $(\vec{k}\vec{v}) = 0$. Vibrations of the fluid in the wave are normal to the propagation direction. The internal waves are transverse ones.

The second parenthesized term in dispersion equation (4.17) is to allow for magnetic field disturbances in the wave. In noninductive approximation $G \gg K_\gamma(1 + \chi_r)$ or when the wave propagation is normal to the field ($\varphi_H = \pi/2$; tg $\varphi_H = \infty$) this term disappears and the internal magnetic wave frequency is determined by the relation $\omega_M^2 = (\mu_0 K G\gamma/\rho)\sin^2\varphi_G$ equivalent to the Brandt-Viyasyaly relation for the internal gravitational waves: $\omega_g^2 = \beta_\rho g\gamma \sin^2 \varphi_g$ (φ_g is the angle between k and g). The internal magnetic waves are low-frequency ones, $\omega_M \sim 1$ Hz. In accordance with (4.17), they do not propagate along the field and temperature gradients when $\varphi_g = 0$.

The group velocity of waves is related as $\vec{V} = \partial\omega/\partial\vec{k}$. By directing the axis y along the field intensity gradient $\vec{G}$ and the axis × normal to it so that the vector k lies in the plane $x0y$ and $\sin \varphi G = K_x/k$, we obtain

$$V_x = \partial\omega/\partial\kappa_x = \omega_0\kappa_y^2/\kappa^3;$$
$$V_y = \partial\omega/\partial\kappa_y = -\omega_0\kappa_x\kappa_y/\kappa^3$$

where $\omega_0 = \sqrt{\mu_0 KG\gamma/\rho}$. If $k_y = 0$, the group velocity of these waves is zero, i.e., they do not propagate in the direction normal to the field gradient either.

With the gravitational field, the gravitational mechanism of internal wave propagation is actuated and, consequently in such a combination these waves may be referred to as magnetogravitational ones. Their frequency is specified by the relation

$$\omega_{Mg}^2 = (\beta g + \mu_0 KG/\rho)\gamma\sin^2\varphi_\gamma$$

where φ_γ is the angle between $\vec{k}$ and $\vec{\gamma}$. It is assumed that the gravitational acceleration is parallel to the field gradient and antiparallel to the temperature gradient ($\vec{g} \uparrow\uparrow \vec{G} \uparrow\downarrow \vec{\gamma}$).

The participation of viscosity and heat conduction leads to attenuation of internal waves. In this case, the dispersion equation is of the form

$$(\nu\kappa^2 - i\omega)(\kappa\kappa^2 - i\omega) + \omega_{Mg}^2 = 0 \tag{4.18}$$

The imaginary part of the wave number corresponds to spatial wave attenuation and may be determined from (4.18).

It follows from dispersion equation (4.17) that the internal magnetic waves may also propagate in a uniform magnetic field when $G = 0$. Then their frequency is

$$\omega_M^2 = \frac{\mu_0 K^2\gamma^2}{\rho(1+\chi_r)} \cdot \frac{\sin^2\varphi_\gamma}{1 + A\ \mathrm{tg}^2\varphi_H} \tag{4.19}$$

and possesses pronounced anisotropy relative to the field direction.

With the existence of the magnetocaloric effect, it appears that internal magnetic waves may also exist in an isothermal fluid. The term $\vec{v}\vec{\gamma}$ in the equation for temperature is substituted by the expression describing the magnetocaloric effect. The wave propagation mechanism is, in this case, due to the fact that the travel of a fluid volume element in the nonuniform field changes its temperature and, hence, magnetization. This results in the restoring force which causes the element to vibrate at the equilibrium position. The dispersion wave equation here is of the form similar to (4.17):

$$\omega_M^2 = \frac{(\mu_0 KG)^2 T^*}{\rho^2 C_{\rho H}} \sin^2\varphi_G \tag{4.20}$$

The magnetocaloric effect is also a mechanism which allows the magnetic field energy to be supplied into the internal wave. Here, if the frequency of the external magnetic field variation coincides with the internal wave frequency, the wave's resonance oscillation increases.

CHAPTER

FIVE

SURFACE PHENOMENA

Another vast region of research interest encompasses specific MF properties associated with thermomechanical processes promoted by the free surface of the magnetic fluid. Being responsible for the form of the free surface, the magnetic field greatly affects the dynamics of film flows. Besides, the magnetic field causes various instabilities and resonance phenomena to take place. On the whole, the magnetic field is an efficient tool used in controlling dynamic processes on the free MF surface. The control is implemented thanks to two factors, namely, the volume magnetic force and magnetic pressure jump on the surface. Here distortions of the magnetic field at the interface are essential. Mathematically, these problems feature the necessity to satisfy the conjugate boundary conditions for the magnetic field on the surface the form of which is not defined in many cases and is an unknown quantity itself.

5.1 FILM FLOWS

In this section, we shall analyze the magnetic field and the film flow it causes in fluids. With essential predominence of magnetic forces over gravitational ones, the latter may be completely ignored thus providing for operation of the film apparatus at any orientation both in space and weightless environments. Comparative simplicity of establishing magnetic fields of various spatial configurations results in a diversity of stable spatial MF films and their flows. Gravitational analogy may as well be used in these problems provided the action of the magnetic force is replaced by some effective gravitational field.

Plane-Parallel Film Flow Along the Field Gradient

A film flow of normal liquid is known to take place on an inclined surface by gravity. Let us consider it for a magnetic fluid when a nonuniform magnetic field exists. The problem geometry is shown in Fig. 5.1 where g_x, $\vec{G}_\tau = [G_x, G_y]$ are, respectively the projection of the gravitational acceleration and the field gradient onto the flow plane. They induce a plane-parallel $\vec{v} = [v_x(z), v_y(z)]$ film flow of a viscous magnetic fluid along the solid surface. The film thickness is l, its surface coordinate $z = 0$. The solid plane coordinte $z = -l$, the flow velocity on it vanishes. The film is assumed to be in a gas flow; their interaction being approximated by a constant shear stress vector $\vec{\tau}$ on the free surface. The latter is related by the boundary conditions at $z = 0$.

$$\eta(\partial v_x/\partial z) = \tau_x, \ \eta(\partial v_y/\partial z) = \tau_y$$

The free film flow under consideration is described by equations (2.34) being for the moment of the form

$$\eta(\partial^2 v_x/\partial z^2) = -\rho g_x - \mu_0 M G_x; \ \eta(\partial^2 v_y/\partial z^2) = -\mu_0 M G_y$$

and yielding the following velocity profile in the film at constant values of gravitational and magnetic forces:

$$\left.\begin{aligned} v_x &= \frac{g_x + \mu_0 M G_x/\rho}{\nu}(l^2 - z^2) + \frac{\tau_x}{\eta}(z + l); \\ v_y &= \frac{\mu_0 M G_y/\rho}{\nu}(l^2 - z^2) + \frac{\tau_y}{\eta}(z + l) \end{aligned}\right\} \quad (5.1)$$

Velocity distribution in the film is the same as for a normal liquid with the

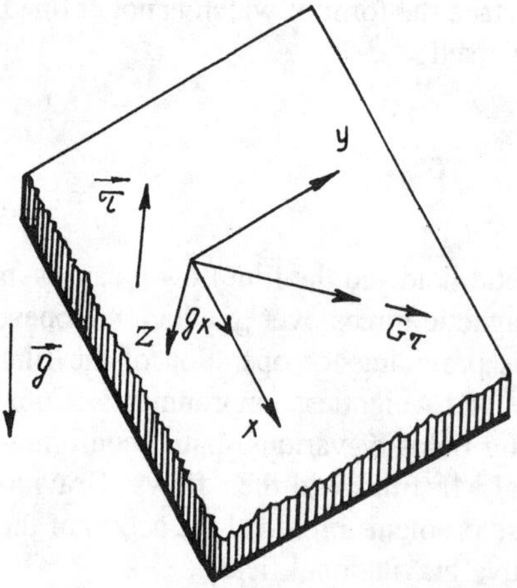

FIG. 5.1

redefined gravitational force $\vec{g} = \vec{g} + \mu_0 MG$. This allows the film flow to be controlled with the help of a magnetic field by regulating the fluid flow rate and film thickness.

With a film in a countercurrent gas flow, the flow is observed to be flooded when the external fluid layers entrain the gas flow. It implies a zero fluid flow rate $\int_0^l v dz = 0$ at the prescribed film thickness. For a cocurrent fluid and gas flow, the condition of "flooding" onset is:

$$\tau \geqslant (2/9)(\rho g_\tau + \mu_0 MG_\tau)$$

Hence, it follows that the magnetic field hinders the flow "flooding" when the intensity gradient direction coincides with the gravitational force and promotes it when the direction is opposite.

Nonisothermal Film Flow on a Cylinder

Here we shall consider a film flow normal to the volume force due to the difference in the fluid levels relative to the equilibrium one. Besides, we shall assume that the surface tension coefficient and fluid magnetization are variable in space, in particular, because of nonuniform temperature distribution.

Let us consider a steady film flow of a weightless MF along a cylindrical (radius R) conductor with current I inducing an axially symmetric and radially nonuniform magnetic field $H_\varphi(r) = I/(2\pi r)$ (Fig. 5.2). The free surface of the film has an axial symmetry and its coordinate is $\xi(z)$. It is assumed that the free surface curvature in longitudinal cross-section is so small that the radial fluid flow and the only nonzero axial velocity component v_z vs z may be ignored. The same condition allows magnetization M and surface tension coefficient α to be regarded as the prescribed function of the coordinate z alone.

The problem is described by the ferrohydrodynamic equations having for this particular case the form

$$\frac{\partial P}{\partial z} = \eta \frac{1}{r} \frac{\partial}{\partial r}\left(r \frac{dv}{dr}\right) \tag{5.2}$$

$$\partial P/\partial r = \mu_0 M \,(dH/dr) \tag{5.3}$$

The gas pressure, P_g, above the fluid surface being set constant, for the pressure jump on it at $r = \xi$ we have

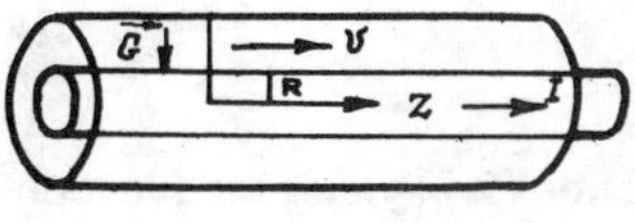

FIG. 5.2

$$P = P_g + \alpha(1/R_1 + 1/R_2)$$

Taking account of all this in the above problem statement, equation (5.3) gives for pressure distribution in the fluid

$$P = P_g + \alpha/\xi + \mu_0 M[H(r) - H(\xi)] = P_g + \alpha/\xi + \frac{\mu_0 MI}{2\pi}\left(\frac{1}{r} - \frac{1}{\xi}\right)$$

With regard to this relation we can rewrite equation (5.2) for velocity

$$\frac{1}{r}\frac{d}{dr}\left(r\frac{dv}{dr}\right) = \frac{\mu_0 I}{2\pi\eta}\frac{dM}{dz}\frac{1}{r} - \frac{1}{\eta}\frac{d}{dz}\left[\left(\frac{\mu_0 MI}{2\pi} - \alpha\right)\frac{1}{\xi}\right] \tag{5.4}$$

The boundary conditions for velocity imply adhesion to the solid wall and shear stress balance on the free one

$$v = 0 \text{ at } r = R;\ \eta(dv/dr) = d\alpha/dz \text{ at } r = \xi$$

Besides, the condition for constant fluid flow rate

$$2\pi\int_R^{\xi} r\,v\,dr = Q \tag{5.5}$$

determines the form of the free surface.

The solution to the problem is

$$v = \frac{\mu_0 I}{2\pi\eta}\left[\frac{dM}{dz}\left(r - R - \xi\ln\frac{r}{R}\right) - \frac{1}{4}\frac{d}{dz}\left(\frac{M - M_\alpha}{\xi}\right)\left(r^3 - R^2 - 2\xi^2\ln\frac{r}{R}\right) + \frac{dM_\alpha}{dz}\xi\ln\frac{r}{R}\right] \tag{5.6}$$

Here $M_\alpha = 2\pi\alpha/(\mu_0 I)$ is introduced. Note that at $M_\alpha > M$ the cylindrical fluid column becomes unstable (cf. 5.5), and $M_\alpha < M$. is set in what follows.

From (5.5) and using (5.6) we have

$$\frac{\eta Q}{\mu_0 I} = \frac{dM}{dz}\left(\frac{7}{12}\xi^3 - \frac{1}{2}\xi^3\ln\frac{\xi}{R} - \frac{1}{2}R\xi^2 - \frac{1}{4}R^2\xi + \frac{1}{6}R^3\right) - \frac{1}{4}\frac{d}{dz}\left(\frac{M - M_\alpha}{\xi}\right)\left(\frac{3}{4}\xi^4 - \xi^4\ln\frac{\xi}{R} - R^2\xi^2 + \frac{1}{4}R^4\right) + \frac{1}{4}\frac{dM_\alpha}{dz}\xi\left(2\xi^2\ln\frac{\xi}{R} - \xi^2 + R^2\right) \tag{5.7}$$

Solution of the ordinary differential first-order equation (5.7) yields the function

$\xi(z)$ to specify the free surface form. Here, the functions $M(z)$ and $M_\alpha(z)$ are assumed to be known. Using (5.7), we can eliminate the derivative $d\xi/dz$ from relation (5.6) for the velocity profile. At the same time, in order to facilitate analysis and numerical calculations it is reasonable that we introduce the dimensionless coordinates $r_1 = r/R$, $\zeta = \xi/R$ (r_1 ranging from 1 to ζ). Then

$$v = \frac{\mu_0 I(dM/dz)R}{2\pi\eta}\Bigg\{ r_1 - 1 + (\sigma_M - 1)\zeta \ln r_1 + \\ + \frac{r_1^2 - 1 - 2\zeta^2 \ln r_1}{(3/4)\zeta^4 - \zeta^4 \ln\zeta - \zeta^2 + 1/4}\Big[A_Q - \frac{1}{2}(\sigma_M - 1)\zeta^3 \ln\zeta + \\ + \frac{1}{4}(\sigma_M - \frac{7}{3})\zeta^3 + \frac{1}{2}\zeta^2 - \frac{1}{4}(\sigma_M - 1)\zeta - \frac{1}{6}\Big]\Bigg\} \tag{5.8}$$

Here

$$\sigma_M = \frac{\partial M_\alpha/\partial z}{\partial M/\partial z}; \quad A_Q = \frac{\eta Q}{\mu_0 I(dM/dz)R^3}$$

The dimensionless velocity profile $v_l = 2\pi\eta\, v/[\mu_0 I(dM/dz)R]$ calculated by this formula for $\zeta = 2$ and $\sigma_M = 1$ is presented in Fig. 5.3 for various fluid flow rates Q (different values of the parameter A_Q). In this case ($\sigma_M = 1$), and the thermocapillary (dM_α/dz) and thermomagnetic (dM/dz) mechanisms contribute to initiation of motion. The thermocapillary mechanism disappears at $\sigma_M = 0$ ($dM_\alpha/dz = 0$), while the thermomagnetic one at $\sigma_M = \infty$ ($dM/dz = 0$). The velocity profiles V_1 and $V_2 = 2\pi\eta v/[\mu_0 I(dM_\alpha/dz)R]$, corresponding to these cases, are shown in Fig. 5.4 at $\zeta = 2$ and zero fluid flow rate

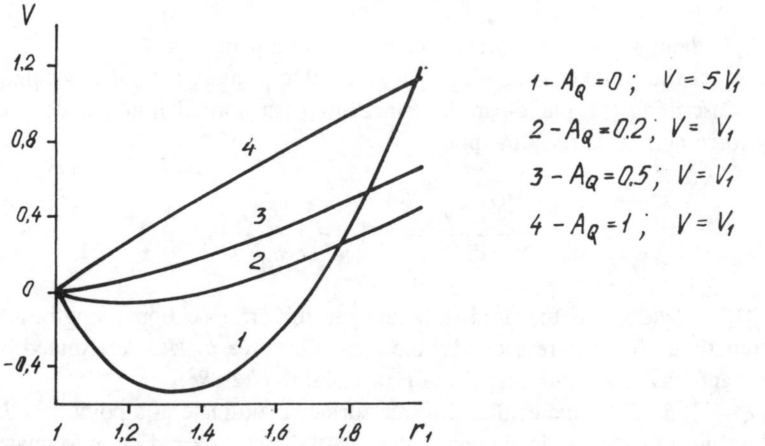

FIG. 5.3

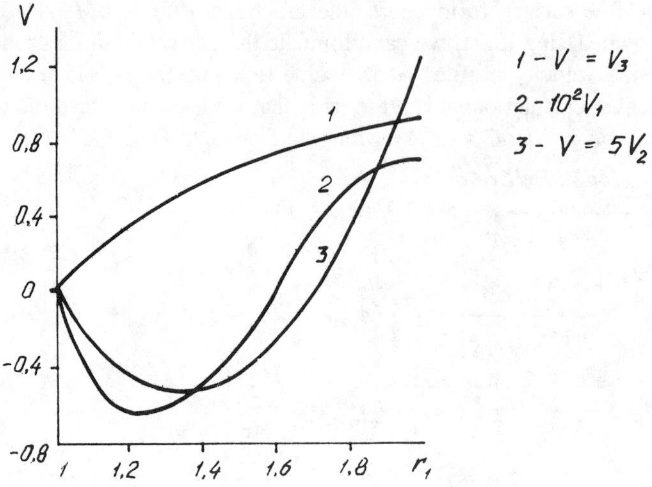

FIG. 5.4

($A_Q = 0$). The velocity profile $V_3 = 2\pi R^2 v/Q$ is for an isothermal fluid ($dM/dz = dM_\alpha/dz = 0$).

In an isothermal fluid, the difference in the film surface levels along the flow is the only moving mechanism, and the equation for the free surface from (5.7) is

$$z = \frac{\mu_0 MIR^3}{4\eta Q}\left[\frac{13}{36}(\zeta^3 - \zeta_0^3) - \frac{1}{3}(\zeta^3 \ln \zeta - \zeta_0^3 \ln \zeta_0) - (\zeta - \zeta_0) - \frac{1}{4}\left(\frac{1}{\zeta} - \frac{1}{\zeta_0}\right)\right] \tag{5.9}$$

where ζ_0 is the dimensionless surface coordinate at a point $z = 0$.

At small film thickness ($\zeta - 1 \equiv \delta \ll 1$), this formula naturally becomes the one describing the plane-parallel film flow of a normal liquid under the gravity force normal to the flow plane

$$z = \frac{\mu_0 MG\,(\xi_0 - R)^4}{12\eta Q_1}\left[1 - \left(\frac{\delta}{\delta_0}\right)^4\right]$$

where $Q_1 = Q/(2\pi R)$ is the fluid flow rate per unit cross-sectional length, and the gravitational force is replaced by the magnetic force $\mu_0 MG$ determined by the constant magnetic field gradient G now equal to $I/(2\pi R^2)$.

If $\zeta = \zeta_L$ is the value of the film free surface coordinate at a point $z = L$, then the fluid flow rate Q is expressed through the levels ζ_0 and ζ_L as follows:

$$Q = \frac{\mu_0 MIR^3}{4\eta L} \left[\frac{13}{36}(\zeta_L^3 - \zeta_0^3) - \frac{1}{3}(\zeta_L^3 \ln\zeta_L - \zeta_0^3 \ln\zeta_0) - (\zeta_L - \zeta_0) - \frac{1}{4}\left(\frac{1}{\zeta_L} - \frac{1}{\zeta_0}\right)\right]$$

and at small film thickness

$$Q_1 = \frac{\mu_0 MG(\xi_0 - R)^4}{12\eta L}\left[1 - \left(\frac{\delta_L}{\delta_0}\right)^4\right]$$

For zero flow rate ($Q = 0$) of a nonisothermal fluid at small film thickness ($\delta \ll 1$), equation (5.7) takes on the form

$$\frac{M - M_\alpha}{3}\frac{d\delta^2}{dz} + \frac{1}{4}\frac{dM}{dz}\delta^2 = \frac{dM_\alpha}{dz}$$

and has the solution in quadratures

$$\delta^2 = \frac{3}{\psi_M}\left[\int \frac{\psi_M}{M - M_\alpha} dM_\alpha + \text{const}\right] \tag{5.10}$$

where

$$\psi_M = \exp\left[\frac{3}{4}\int \frac{dM}{M - M_\alpha}\right]$$

It is peculiar of the situation that, even in a case of a thin film, the constant curvature in the cross-section is represented by the difference $M - M_\alpha$ in the formulae for the free surface profile. And only at $M \gg M_\alpha$ does (5.10) become a relation for a plane-parallel flow

$$\mu_0 G(R\delta)^2 = 3M^{-3/4}\left[\int M^{-1/4} d\alpha + \text{const}\right]$$

5.2 SURFACE INSTABILITY

Instability of MF surfaces in the magnetic field normal to them is one of the striking features of magnetic fluids. This instability manifests itself when the magnetic field attains a certain critical value, to be more exact, a certain value of fluid magnetization is achieved, the fluid surface is not smooth any longer and cone-like peaks spasmodically appear on its surface, with geometric characteristics determined by the magnetic field configuration and its intensity.

The physical mechanism of this instability is attributed to the fact that any perturbations on the free fluid surface lead to magnetic field distortions which cause their further growth. The magnetic pressure jump also plays a certain part.

For example, if a hump-like perturbation appears on a plane surface, then, because of the pressure jump, distribution in the hump becomes minimal at the top where the normal magnetization component has the highest value. The situation is equivalent to that considered in 3.3 for a liquid droplet. In order to balance this differential pressure, still greater surface curvature at the hump top, i.e., its development, is required. The insufficient pressure in the depression in the surface is balanced by the gravitational hydrostatic pressure drop between the hump top and the depression. The surface becomes wavy, with more pronounced curvature at hump tops than in depressions. The form of the surface becomes stationary due to the equilibrium of magnetic, gravitational and capillary forces. If the magnetic field is nonuniform, then within the gravitational analogy the equilibrium magnetic force is as important as the gravitational force, which is substituted by the effective quantity

$$\rho \vec{g}_{\text{eff}} = \rho \vec{g} + \mu_0 M \nabla H$$

The threshold instability characteristics are estimated within the linearized theory. Equilibrium quantities will have subscript 0; perturbations will be primed. Moreover, quantities for the magnetic fluid layer, the media above and below it will correspondingly have subscripts 1, 2 and 3. For small perturbations $\vec{v}\,' = \vec{v} - \vec{v}_0$, $\vec{h} = \vec{H} - \vec{H}_0 = \nabla\psi$, $\vec{m} = \vec{M} - \vec{M}_0$, $H' = H - H_0$, $M' = M - M_0$, $P' = P - P_0$ the ferrohydrodynamics equations give in the linear approximation

$$\left.\begin{aligned} &\rho\left[\frac{\partial \vec{v}}{\partial t} + (\vec{v}_0 \nabla)\vec{v}\,' + (\vec{v}\nabla)\vec{v}_0\right] = -\nabla P' + \eta\Delta\vec{v}\,' + \mu_0 \chi_s (\vec{H}_0 \nabla)\vec{h} - \\ &- \mu_0 \mu_s \sigma_\chi \frac{(\vec{H}_0 \vec{h})}{H_0} \nabla H_0 ; \\ &\operatorname{div} \vec{v} = 0 ; \ \Delta\psi = \sigma_\chi (\vec{e}\nabla)^2 \psi \end{aligned}\right\} \quad (5.11)$$

Here

$$H' = (\vec{H}_0 \vec{h})/H_0; \ M' = (\vec{M}_0 \vec{m})/M_0 = \chi_r H', \ \chi_r = \partial M_0/\partial H_0 ;$$

$$\vec{m} = (M_0/H_0)\vec{h} - (\chi_s - \chi_r)\frac{(\vec{H}_0 \vec{h})}{H_0}\frac{\vec{H}_0}{H_0} ; \ \chi_s = M_0/H_0 ;$$

$$\vec{e} = \vec{H}_0/H_0; \ \sigma_\chi = (\chi_s - \chi_r)/(1 + \chi_s); \ \mu_r = 1 + \chi_r ; \ \mu_s = 1 + \chi_s$$

In linearizing the interface boundary conditions for plane fluid layers (Fig. 5.5) we should remember that the vector of the normal to the surface for surface perturbations, prescribed in the form $z = \zeta(t,x,y)$, has components $\vec{n} = [-\partial\xi/\partial x, -\partial\xi/\partial y\ 1]$, and the surface curvature is $(R_1^{-1} + R_2^{-1}) = -\Delta\xi$. Again, when estimating zero-approximation values on the perturbed sur-

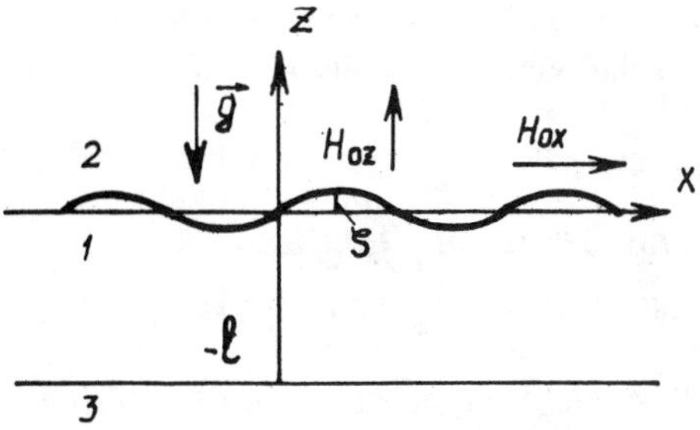

FIG. 5.5

face, one should take account of their possible spatial nonuniformity and use relations that are the expansion of functions in terms of the direction of the normal to the surface of the type $\vec{v}_0(\xi) = \vec{v}_0(0) + [(\vec{n}_0\nabla)\vec{v}_0]\xi$ for vector quantities and $P_0(\xi) = P_0(0) + (\vec{n}_0\nabla P_0)\xi$ for scalar quantities.

Below, the difference of values to both sides of the interface will be bracketed as, for example, $\{P\} \equiv P_1(\xi) - P_2(\xi)$.

In view of the above, boundary conditions (2.48) and (2.41) on the perturbed interface of two MFs will be written in the linear approximation as

$$\{\eta\,(\partial v_x'/\partial z + \partial v_z'/\partial x) + \xi\,(\partial^2 v_{0x}/\partial z^2)\} = 0;$$
$$\{\eta\,(\partial v_y'/\partial z + \partial v_z'/\partial y) + \xi\,(\partial^2 v_{0y}/\partial z^2)\} = 0 \tag{5.12}$$

$$\{2\eta\,(\partial v_z'/\partial z\} - \{P'\} - a\Delta\xi - \{\partial P_0/\partial z + \mu_0\chi_s^2 H_{0z}\,(\partial H_{0z}/\partial z\}\,\xi -$$
$$- \mu_0\{\chi_s^2 H_{0z}H_{0x}\}\ \partial\xi/\partial x - \mu_0\{\chi_s^2 H_{0z}H_{0y}\}\ \partial\xi/\partial y -$$
$$- \mu_0\{\chi_s^2 H_{0z}\ [(1 - e_z^2\,(\sigma_\chi\mu_s/\chi_s))h_z - e_z e_x h_x\,(\sigma_\chi\mu_s/\chi_s) -$$
$$- e_z e_y h_y\ (\sigma_\chi\mu_s/\chi_s)]\} = 0$$

$$\{\partial B_{0z}/\partial z\}\ \xi - \{\mu_s H_{0x}\}\ \partial\xi/\partial x - \{\mu_s H_{0y}\}\ \partial\xi/\partial y +$$
$$+ \{(\mu_s - \sigma_\chi\mu_s e_z^2)h_z\} - \{\sigma_\chi\mu_s e_z e_x h_x\} - \{\sigma_\chi\mu_s e_z e_y h_y\} = 0; \tag{5.13}$$

$$\{\partial H_{0y}/\partial z\}\ \xi + \{H_{0z}\}\ \ \partial\xi/\partial y + \{h_y\} = 0;$$
$$\{\partial H_{0x}/\partial z\}\ \xi + \{H_{0z}\}\ \partial\xi/\partial x + \{h_x\} = 0$$

$$\partial\xi/\partial t + \vec{v}_0\nabla\xi = v_z' \quad \text{– for both media} \tag{5.14}$$

In the linear approximation, these boundary conditions must be satisfied at an equilibrium surface, for example $z = 0$. On a solid boundary, the viscous fluid velocity is zero, and for an ideal fluid the nonflow condition $(\vec{v}\vec{n} = 0$ must hold.

The Cauchy-Lagrange Integral and Bernoulli Equation

In an isothermal fluid, the volume magnetic force $\vec{f}_M = \mu_0 M \nabla H$ has the potential $\Phi = \mu_0 \int_0^H MdH$. Owing to this, the equation of motion for MFs includes the same integrals as for a normal liquid. The Cauchy-Legrange integral for a potential flow $(\vec{v} = \nabla\varphi)$ of an ideal incompressible fluid is of the form

$$P + \rho(\partial\varphi/\partial t) + 0{,}5\rho(\nabla\varphi)^2 + \rho g z - \mu_0 \int_0^H MdH = \text{const} \tag{5.15}$$

The continuity equation is this case takes on the form of the Laplace equation for the velocity potential

$$\Delta\varphi = 0 \tag{5.16}$$

For a steady vortex flow, the Bernoulli integral holds along the streamline

$$P + 0{,}5\rho v^2 + \rho g z - \mu_0 \int_0^H MdH = \text{const} \tag{5.17}$$

Linear pressure perturbations in a fluid are found from the Cauchy-Lagrange integral as follows:

$$P' = -\rho(\partial\varphi'/\partial t) - \rho(\vec{v}_0\nabla\varphi') + \mu_0 M_0(\vec{H}_0\vec{h})/H_0 \tag{5.18}$$

while the equilibrium pressure distribution is related by

$$\nabla P_0 = \rho\vec{g} + \mu_0 M_0 \nabla H_0$$

Wave Solutions for Velocity and Field Potentials

The stability of magnetic fluid layers will be studied relative to plane wave-type perturbations

$$\xi = \xi_a \exp i(\vec{\kappa}\vec{r} - \omega t) \tag{5.19}$$

propagating on the interface $\vec{\kappa} = [\kappa_x, \kappa_y, 0]$. For such perturbations the Laplace equation for velocity potential (5.16) has a general solution of the form

$$\varphi' = (C_1 e^{\kappa z} + C_2 e^{-\kappa z})\exp i(\vec{\kappa}\vec{r} - \omega t) \tag{5.20}$$

while the solution to Maxwell equations in system (5.11) determines the potential of magnetic field perturbations ($\vec{h} = \nabla\psi$)

$$\psi = [C_3 \exp(\kappa_R + i\kappa_I)z + C_4 \exp(-\kappa_R + i\kappa_I)z] \exp i(\vec{\kappa}\vec{r} - \omega t)$$

where

$$\kappa_R = \left[\frac{\kappa^2 - \sigma_\chi e_z^2 \kappa^2 - \sigma_\chi e_x^2 \kappa_x^2 - \sigma_\chi e_y^2 \kappa_y^2 - 2\sigma_\chi^2 e_z^2 e_x e_y \kappa_x \kappa_y}{(1 - \sigma_\chi e_z^2)^2}\right]^{1/2} \tag{5.21}$$

$$\kappa_I = \frac{\sigma_\chi e_z (e_x \kappa_x + e_y \kappa_y)}{1 - \sigma_\chi e_z^2}$$

provided σ_x, e_x, e_y, e_z = const.

The potentials of velocity and magnetic field perturbations in unbounded media should also meet the condition of their attenuation at infinity ($z = \pm\infty$). They are therefore specified by one of the terms, $e^{-\kappa z}$ or $e^{\kappa z}$, in (5.20) and (5.21), depending on a half-space (positive or negative) these unbounded media are in.

Substituting these solutions into the boundary conditions, we obtain, in particular, a set of linear uniform algebraic equations based on amplitude constants. The requirement that the determinant of this set vanish provides the dispersion equation $D(\omega, \kappa)$ the analysis of which leads to the stability conditions for the set relative to the above perturbations. The roots ω of this dispersion equation that possess a positive imaginary part $\mathrm{Im}\omega > 0$ point to instability accompanied by an exponential growth of the perturbations in time.

$$\xi \sim \exp[\mathrm{Im}(\omega)t]$$

With this in view, the effect of different factors on the stability of MF films and film flows in magnetic fields of various spatial configuration is discussed.

Stability of Thick MF Layer Surface

The simplest situation is realized when a fixed ($\nu_0 = 0$) layer of ideal ($\eta = 0$) magnetic fluid is unbounded ($l = \infty$) inward from the free surface, with a nonmagnetic gas ($\chi_2 = 0$, $\rho_2 = 0$) above it (cf. Fig. 5.5). A uniform magnetic field has only one vertical component $\vec{H}_0 = 0, 0, H_0]$. $\vec{e} = [0, 0, 1]$ normal to an equilibrium surface $z = 0$. The dispersion equation for surface waves in this case is of the form

$$\omega = \pm \frac{\kappa}{\sqrt{\rho}} \sqrt{\frac{\rho g + \alpha\kappa^2}{\kappa} - \frac{\mu_0 M_0^2}{1+(\mu_r \mu_s)^{-1/2}}} \tag{5.22}$$

The neutral stability curve is specified by the condition $\omega = 0$, i.e., instability is monotonic and supercritical equilibrium is steady.

The surface instability condition following from (5.22) implies

$$\mu_0 M_0^2 \geqslant \left(1 + \frac{1}{\sqrt{\mu_r \mu_s}}\right) \frac{\rho g + \alpha \kappa^2}{\kappa} \tag{5.23}$$

Minimization of the wave number κ in the rhs of (5.23) yields critical (threshold) values (with subscript$_*$)

$$\mu_0 M_*^2 = 2\sqrt{\rho g \alpha}\left(1 + \frac{1}{\sqrt{\mu_r \mu_s}}\right); \quad \kappa_* = \sqrt{\rho g / \alpha} \tag{5.24}$$

At a magnetic field intensity high enough to ensure fluid magnetization above critical, the fluid surface becomes unstable. The surface perturbation wavelength is determined by a capillary radius $\sqrt{\alpha/\rho g}$. Numerical estimates show that the critical magnetization values M_* are of the order of 10^4 A/m, while the critical wavelength $\Lambda_* = 2\pi/\kappa \sim 1$ cm. From (5.24), it follows, in particular, that surface instability cannot be observed in any magnetic fluid. It will not take place, for example, if saturation magnetization is less than critical.

The surface instability may approximately be analyzed using dimensionless quantities. For this, let us introduce the dimensionless wave number $S = \kappa\sqrt{\alpha/\rho g}$, the Bond number $\text{Bo} = \rho g l^2/\alpha$ and the surface instability criterion $\text{Si} = \mu_0 M^2/\sqrt{\rho g \alpha}$. For the case we shall have the equation for the neutral stability curve in the form

$$\text{Si} = \frac{1 + S^2}{S}\left(1 + \frac{1}{\sqrt{\mu_r \mu_s}}\right) \tag{5.25}$$

The concept "big layer thickness" is determined by the condition $\text{Bo} \gg 1$, i.e., the layer is much thicker than the capillary radius $l \gg \sqrt{\alpha/\rho g}$ or longer than the wavelength of developing perturbations.

If there is a nonmagnetic fluid, whose density cannot be ignored, above the MF surface, then the capillary radius and the instability surface criterion are determined in terms of the fluid density difference $\Delta\rho = \rho_1 - \rho_2$: $S = \kappa\sqrt{\alpha/\Delta\rho g}$, $\text{Si} = \mu_0 M_0^2/\sqrt{\Delta\rho g \alpha}$. In the situation considered, relations (5.25) hold. The denominate wavelength $\Lambda = 2\pi/\kappa$ here increases: $\Lambda_* = 2\pi\sqrt{\alpha}/\sqrt{\Delta\rho g}$, while critical magnetization decreases:

$$\mu_0 M_*^2 = 2\sqrt{\Delta\rho g \alpha}\,[1 + (\mu_r \mu_s)^{-1/2}]$$

In analyzing the stability of the interface of two magnetic fluids, Si is defined as a squared difference of their magnetizations: $\text{Si} = \mu_0(M_{01} - M_{02})^2/\sqrt{(\rho_1 - \rho_2)g\alpha}$. In this case, for infinitely thick layers we have

$$\text{Si}_* = 2\left(\frac{1}{\sqrt{\mu_{r1}\mu_{s1}}} + \frac{1}{\sqrt{\mu_{r2}\mu_{s2}}}\right), \quad S_* = 1 \tag{5.26}$$

Instability of Finite-Thickness Plane Layer Surface

If a plane MF layer is l in thickness and confined from below by a solid surface, the neutral curve for its surface stability will be

$$\mathrm{Si} = \frac{1+S^2}{S} \cdot \frac{1+\beta_{23}+(\beta_{13}+\beta_{21})\,\mathrm{tg}(\gamma_1\sqrt{\mathrm{Bo}}\,S)}{\beta_2[1+\beta_{13}\,\mathrm{th}(\gamma_1\sqrt{\mathrm{Bo}}\,S)]}^{*} \tag{5.27}$$

where

$$\gamma_1 = \sqrt{\mu_{s1}/\mu_{r1}}\,,\ \beta_i = \sqrt{\mu_{ri}\mu_{si}}\,,\ \beta_{ij} = \beta_i/\beta_j$$

For a thick layer (Bo $\gg$ 1), this formula gives relations (5.25) and (5.26).

For a thin layer (Bo $\ll$ 1), equation (5.27) yields the following threshold values:

$$\mathrm{Si}_* = 2(\beta_2^{-1}+\beta_3^{-1}) = 2\left(\frac{1}{\sqrt{\mu_{r2}\,\mu_{s2}}} + \frac{1}{\sqrt{\mu_{r3}\mu_{s3}}}\right), S_* = 1 \tag{5.28}$$

The critical wave number for thin and thick layers is unity. The critical value of the surface instability criterion is only determined by the magnetic characteristics of the media bounding the thin layer (β_2, β_3). So, if the layer is confined from above and from below by nonmagnetic media ($\beta_2 = \beta_3 = 1$), then $\mathrm{Si}_* = 4$. If the upper medium is nonmagnetic ($\beta_2 = 1$), and the lower one has high magnetic permeability ($\beta_3 \gg 1$), then $\mathrm{Si}_* = 2$. Thus, a medium of high magnetic permeability confining the thin layer from below de-stabilizes it.

In the above sense, the thin layer is a layer about 1 mm thick.

Total dependence of the critical values of the surface instability criterion and wave number on the Bond number, calculated by formula (5.27), is plotted in Fig. 5.6 for a fluid obeying the linear magnetization law $\gamma_1 = 1$, $\beta_1 = 2$, confined by nonmagnetic fluid $\beta_2 = 1$ from above and by nonmagnetic solid boundary ($\beta_3 = 1$), curve 1, and high magnetic permeability boundary ($\beta_3 = 1000$), curve 2, from below. The effect of the lower boundary is felt to about Bond number 2.

Surface Instability of a Plane Layer With Two Free Boundaries

When a thin MF layer is floating, for example, on a heavier fluid and is adjacent to air from above, its both boundaries are deformable (free). In the general case, its stability condition is rather tedious, and we shall, therefore, consider a situation when the density differences of two adjacent media are equal $\rho_3 - \rho_1 = \rho_1 - \rho_2 = \Delta\rho$ and the surface tension coefficients at both interfaces are also the same and equal to α. The neutral stability curve now is of the simple form

* See Nomenclature p. 211 for explanation of notation for trigonometric functions.

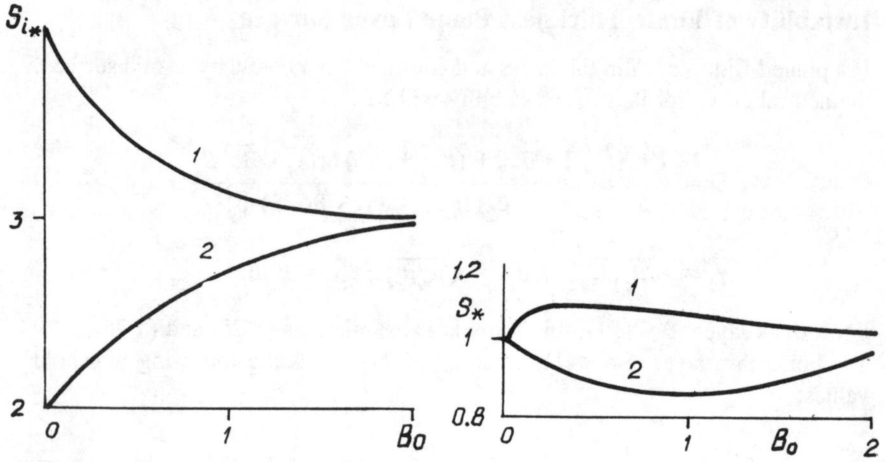

FIG. 5.6

$$\text{Si} = (1/2)\,(\beta_2^{-1} + \beta_3^{-1})\,(1 + S^2)/S\,, \tag{5.29}$$

where

$$\text{Si} = (1/2)\mu_0[\,(M_{01} - M_{02})^2 + (M_{01} - M_{03})^2]/\sqrt{\Delta\rho g a}$$

is determined through the arithmetical mean of squared magnetizations of the layer fluid and bounding media. From (5.29), we obtain the following critical values of the parameters:

$$\text{Si}_* = (\beta_2^{-1} + \beta_3^{-1}),\ S_* = 1 \tag{5.30}$$

The critical value of Si, in this case, is twice as small as for a similar layer confined from one side by a solid surface (cf. formula (5.28)). For example, for non-magnetic media confining the layer ($\beta_2 = \beta_3 = 1$), we have $\text{Si}_* = 2$. The layer with two free surfaces is less stable.

Stability of a Surface Coated With a Thin Elastic Film

When making use of the deformation properties of a free MF surface, it is required that the fluid be isolated from the surrounding medium with, for example, elastic shell. It is, therefore, reasonable, to study wave processes on the fluid surface as well as its stability, taking account of the elastic properties of the coating film. The equation of motion for a free elastic film is

$$\rho_n l_n(\partial^2\xi/\partial t^2) = -D_y\Delta_1\Delta_1\xi + f_m;\ D_y = l_n^3 E/[12(1 - \sigma_y^2)] \tag{5.31}$$

where ρ_n, l_n are correspondingly film density and thickness; D_y film bending strength; E Young' modulus; σ_y Poisson's ratio; $\Delta_1 = \partial^2/\partial x^2 + \partial^2/\partial y^2$ two-dimensional Laplacean; f_m surface density of forces normal to the film, which in case of negligible surface tension forces, is determined by the difference of normal stresses σ_{nn} in the fluids from both sides of the film: $f_m = \sigma_{nn2} - \sigma_{nn1}$ at $z = 0$. The force is found from relations (5.13) and (5.18).

If an elastic film coats a plane interface of two infinitely thick magnetic fluids obeying linear magnetization laws ($\vec{M} = \chi\vec{H}$) and a uniform magnetic field is normal to it, then the neutral curve of its stability is described by the expression

$$\mu_0 (M_{01} - M_{02})^2 = \frac{\mu_1 + \mu_2}{\mu_1 \mu_2} \frac{D_y \kappa^4 + (\rho_1 - \rho_2) g}{\kappa} \tag{5.32}$$

Whence, the critical magnetization values and wave numbers are estimated as:

$$\left.\begin{aligned} \mu_0 (M_{01} - M_{02})^2_* &= \frac{4\sqrt[4]{3}}{3} \frac{\mu_1 + \mu_2}{\mu_1 \mu_2} [(\rho_1 - \rho_2)^3 g^3 D_y]^{1/4}; \\ \kappa_* &= 3^{-1/4} [(\rho_1 - \rho_2) g / D_y]^{1/4} \end{aligned}\right\} \tag{5.33}$$

For a magnetic fluid with $\rho \sim 10^3$ kg/m^3, $M \sim 10^4$ A/m, instability of the surface, coated with high-pressure polyethylene film, will occur at a film thickness of about 0.1 mm. The critical wavelength here will be about 5 mm.

5.3 TANGENTIAL FIELD EFFECT AND HYDRODYNAMIC DISCONTINUITY STABILITY

A so-called tangential hydrodynamic discontinuity, which is unstable in normal fluids, is said to take place in case of relative motion of fluids along their interface. This stability results in ripples on the surface of water reservoirs on a windy day.

Let us consider the effect of a magnetic field on the stability of this discontinuity in MFs for a plane-parallel layer shown in Fig. 5.5. It is assumed that medium 2 above the layer moves along the axis x at constant velocity $u(\vec{v}_0 = [u_0, 0, 0])$. The lower boundary is solid, and the uniform magnetic field $\vec{H}_0$ has components H_{0x}, H_{0y} and H_{0z} at infinity in a solid. Linear magnetization laws $\vec{M} = \chi\vec{H}$ are employed for all the three media.

The dispersion equation for waves at the fluid interface is now of the form

$$u \equiv \frac{\omega}{\kappa} = \frac{\rho_2 u_0 \cos\theta_{\kappa u}}{\rho_1/\text{th}\,\kappa l + \rho_2} \pm \left\{ \frac{(\rho_1 - \rho_2) g + \alpha\kappa^2}{\kappa(\rho_1/\text{th}\,\kappa l + \rho_2)} - \right.$$ *

* See Nomenclature p. 211 for explanation of notation for trigonometric functions.

$$- \frac{\rho_1\rho_2 u_0^2 \cos^2\theta_{\kappa u}}{(\rho_1/\operatorname{th}\kappa l + \rho_2)^2 \operatorname{th}\kappa l} - \frac{\mu_3^2(\mu_2-\mu_1)^2}{\mu_1\mu_2^2} H_{0z}^2 \times$$

$$\times \frac{1}{(\rho_1/\operatorname{th}\kappa l + \rho_2)} \cdot \frac{\mu_3/\mu_1 + \operatorname{th}\kappa l}{1 + \mu_3/\mu_2 + (\mu_1/\mu_2 + \mu_3/\mu_1)\operatorname{th}\kappa l} +$$

$$+ \frac{(\mu_2-\mu_1)^2}{\mu_2} H_{0\tau}^2 \frac{1}{\rho_1/\operatorname{th}\kappa l + \rho_2} \cdot \frac{1 + (\mu_3/\mu_1)\operatorname{th}\kappa l}{1 + \mu_3/\mu_2 + (\mu_1/\mu_2 + \mu_3/\mu_1)\operatorname{th}\kappa l} \times$$

$$\left. \times \cos^2(\theta_{\kappa u} - \theta_{\kappa H}) \right\}^{1/2}, \quad \mu = 1 + \chi \tag{5.34}$$

where $\theta_{\kappa u}$ is the angle between the vectors $\vec{\kappa}$ and $\vec{u}_0$; $\theta_{\kappa H}$ is the angle between the magnetic field intensity component $\vec{H}_{0\tau} = [H_{0x}, H_{0y}, 0]$ tangential to the interface and the wave vector $\vec{\kappa}$; $H_{0\tau}^2 = H_{0x}^2 + H_{0y}^2$. The reversal of sign, from plus to minus, in the radicand (5.34) for phase velocity points to the onset of surface instability.

In the situation under study, the surface stability is affected by three mechanisms. The first one is due to the normal component of fluid magretization and is described, as in 5.2, by the surface instability criterion

$$\mathrm{Si} = \mu_0(M_{0z1} - M_{0z2})^2 \sqrt{(\rho_1 - \rho_2)g\alpha}$$

As will be seen later, the magnetic field intensity component $H_{0\tau}$ tangential to the interface is, on the contrary, a stabilizing mechanism. It is expedient that this mechanism be described by an independent surface stabilization criterion St built similarly to Si but including the difference of tangential components of medium magnetization

$$\mathrm{St} = \mu_0(M_{0\tau 1} - M_{0\tau 2})^2/\sqrt{(\rho_1 - \rho_2)g\alpha}$$

Finally, the tangential discontinuity effect is described by the known Weber number $\mathrm{We} = \bar{\rho}\,\bar{l}\,\vec{u}_0^2/\alpha$ where the capillary radius $\bar{l} = \sqrt{\alpha/(\rho_1 - \rho_2)g}$ should be chosen as a characteristic dimension and $\bar{\rho} = \rho_1\rho_2/(\rho_1 + \rho_2)$ as density. Then for the case

$$\mathrm{We} = u_0^2\rho_1\rho_2/[(\rho_1 + \rho_2)\sqrt{(\rho_1 - \rho_2)g\alpha}]$$

The Bond number and the dimensionless wave number are estimated in the same way as in 5.2:

$$\mathrm{Bo} = (\rho_1 - \rho_2)gl^2/\alpha, \; s = \kappa\sqrt{\alpha/(\rho_1 - \rho_2)g}$$

Tangential Field Effect on Surface Stability

First of all, let us dwell upon the specific features of interface stability for unbounded ($kl \gg 1$, Bo $\gg 1$) steady-state ($u_0 = 0$, We $= 0$) fluid layers in the presence of a tangential magnetic field. For this case we obtain from (5.34) a neutral stability curve in the form

$$\mathrm{Si} = \left(\frac{1}{\mu_1} + \frac{1}{\mu_2}\right)\frac{1 + S^2}{S} + \frac{\mathrm{St}}{\mu_1 \mu_2}\cos^2\theta_{\kappa H} \tag{5.35}$$

The second term in the rhs of (5.35), describing the tangential field effect, increases the Si values corresponding to an unsteady situation (cf. 5.25); the tangential magnetic field increases the surface stability. But this is only true of those perturbations which have a nonzero projection of the wave vector onto the field direction when $\cos^2 \theta_{\kappa H} \neq 0$.

As for perturbations propagating normal to the field ($\theta_{\kappa H} = \pi/2$),neutral curve (5.35) has the same form as with no field and yields the lowest critical Si number. This kind of perturbation is most detrimental. Hence, we may conclude that in this case the tangential magnetic field only promotes the perturbations propagating normal to it rather than increases the instability onset threshold. Just as in the case of convective stability (cf. 4.2), it eliminates degeneration in the direction of most detrimental perturbations thus hindering the development of perturbations that are nonuniform along its direction.

Rolling waves with axes parallel to the field are expected to be a supercritical form of the surface in the presence of a magnetic field tangential to it. This fact can clearly be seen in experiments.

By generalizing the results on convective and surface instability we may conclude that a uniform magnetic field hinders the development of perturbations which distort it, i.e., perturbations nonuniform along its direction. In the situations when these very perturbations are most detrimental because of other reasons, the uniform magnetic field may also increase the stability threshold of the system.

Tangential Magnetic Field Effect on Hydrodynamic Discontinuity Stability

With no magnetic field (Si $=$ St $= 0$), the discontinuity instability condition is of the form

$$\mathrm{We}\cos^2\theta_{\kappa u} = (1 + s^2)/s \tag{5.36}$$

and indicate that the most detrimental perturbations are those propagating along the flow $\theta_{\kappa u} = 0$ for which the critical Weber number is two (We$_* = 2$, $s_* = 1$). Instability now is of wave character.

The uniform magnetic field tangential to the surface leads to the following equation for the neutral curve

$$\mathrm{We}\cos^2\theta_{\kappa u} = \frac{1+s^2}{s} + \frac{\mathrm{St}}{\mu_1+\mu_2}\cos^2\theta_{\kappa H} \tag{5.37}$$

If the magnetic field is normal to the flow, it does not exert any effect on downstream perturbations ($\theta_{\kappa H} = \pi/2$), i.e., it does not effect the discontinuity stability ($\mathrm{We}_* = 2$). When it is parallel to the flow ($\vec{H}_{0\tau} \uparrow\uparrow \vec{u}_0$, $\theta^*_{\kappa u} = \theta_{\kappa H}$), the critical Weber number will be higher:

$$\mathrm{We}_* = 2 + \mathrm{St}/(\mu_1 + \mu_2),\ s_* = 1 \tag{5.38}$$

The tangential uniform magnetic field parallel to the flow stabilizes the hydrodynamic discontinuity and, at a prescribed velocity, may assume its stability.

In case of intermediate field directions, the critical value of the Weber number is

$$2 < \mathrm{We}_* < 2 + \mathrm{St}/(\mu_1 + \mu_2)$$

In case of parallel vectors $\vec{H}_{0\tau}$ and $\vec{u}_0$, when the field stabilizing effect is the highest, for a thin layer ($kl \ll 1$, $\mathrm{Bo} \ll 1$) we obtain from (5.34)

$$\mathrm{We}_* = 2 + \frac{\mathrm{St}}{\mu_1+\mu_2}\,\frac{\mu_2+\mu_1}{\mu_2+\mu_3},\quad s_* = 1 \tag{5.39}$$

The comparison of this expression with (5.38) shows that confinement of the layer from below with a solid whose magnetic permeability exceeds that of the fluid ($\mu_3 > \mu_1$), reduces the stabilizing effect of the tangential field making it vanish at $\mu_3 \to \infty$. If $\mu_3 < \mu_1$, the stabilizing effect is enhanced.

Taking account of the above results, we may say that a solid interface with magnetic permeability exceeding that of the fluid, exerts a destabilizing effect on fluid stability, while an interface with lower magnetic permeability stabilizes the layer.

Another fact should be mentioned concerning the stability of thin layers. The notion "free surface instability" cannot be applied to a thin layer as was the case with thick layers because instability of the layer immediately causes its disintegration into separate droplets.

It is also necessary to dwell upon a factor such as the effect of magnetic field nonuniformity on the surface stability. Within the framework of gravitational analogy, the action of the volume magnetic force $\mu_0 M \nabla H$ is equivalent to that of the gravitational force $\rho\vec{g}$ and mathematically results in its substitution by the effective value $(\rho\vec{g})_{\mathrm{eff}} = \rho\vec{g} + \mu_0 M \nabla H$. In practice, this means that if ∇H is directed from the free surface inside the fluid, the effective value of the gravitational force increases, the critical fluid magnetization $\mu_0 M_*^2 \sim (\rho g)_{\text{eff}}$ grows, and the critical perturbation wavelength $\Lambda_* \sim (\rho g)_{\text{eff}}^{-1/2}$ decreases. Contrasting conclusions are possible if the fluid gradient is directed from the free surface outward thus reducing the gravitational field effect.

Film Flow Stability

The above results evidence that a magnetic field is an efficient means of controlling processes on a free MF surface. First of all, we should point out the stabilizing effect of a tangential uniform magnetic field and a nonuniform field with stress gradient directed from the surface inside the fluid. These circumstances may be turned to advantage in controlling the stability of fluid film flows by means of a magnetic field in the problems of film heat and mass exchangers.

For instance, considering the stability of a plane – parallel MF film flow, described by formulae (5.1), on a vertical plane ($g_x = g$, $g_z = 0$) in the presence of a magnetic field whose gradient is directed to the wall inside the fluid ($G_x = G_y = 0$, $G_z = G$)(cf. Fig. 5.1), we shall see that this flow will completely be stabilized by a transverse field gradient if

$$\mu_0 MG > (2/5)\rho g(\tau + \rho g l)l^2/\nu\eta \qquad (5.40)$$

The available experimental data show that a nonuniform magnetic field provides a stable MF film flow at velocities of an incoming gas flow several times greater than the critical values for a normal fluid.

It is quite natural that the magnetic field normal to the film surface destabilizes the film flow while the tangential field enhances its stability by increasing the threshold values of unsteady perturbation wavelengths and the time of their development.

5.4 SURFACE-CONVECTIVE INSTABILITY

As already noted, thermoconvective and surface phenomena in MFs represent two classes of processes characterized by direct interconnection and interdependence of thermomechanic and magnetodynamic states of systems.

The study of the interaction of these phenomena is, no doubt, of great interest in situations where they act concurrently. Such situations occur in nonisothermal MF volumes with a deformable fluid-surrounding medium interface. They may be interpreted as thermoconvective processes in a fluid with deformable surface or as surface phenomena in nonisothermal magnetic fluids. We consider them under the name "surface convective phenomena." Surface-convective instability of mechanical equilibrium states is most attractive. As follows from the above, thermoconvective processes in MFs are described by dimensionless criteria, the basic of them being the gravitational Grashof number Gr for the gravitational mechanism of convection; the magnetic Grashof number Gr_m (magnetic mechanism); the parameter D specifying the contribution of thermal field perturbations to the above processes.

When a nonisothermal fluid has free boundaries, yet another mechanism is involved. It is thermocapillary convection related to the temperature dependence of the surface tension coefficient and described by the Marangoni number Ma.

Free MF surface phenomena are described by the surface instability criterion Si, gravitational and magnetic Bond numbers Bo and Bo_m.

Thermal processes at fluid interfaces are specified by the relations of their thermophysical properties or the Biot number (Bi), with the interface heat transfer obeying the Newton law.

Combined surface-convective processes will be described by the set of the above parameters which, owing to a great number of interrelated physical mechanisms, will have a wide range of manifestations. Deformation of a free surface under the action of thermomechanic and magnetic forces is the most serious reason why these mechanisms are interrelated. This deformation is caused by convective motion in a fluid and by magnetostatic instability. For example, the development of surface instability perturbs the medium interface and disturbs the thermal regime at this interface, i.e., disturbs the temperatures which, in turn, cause convective motion in the fluid. Therefore, the conditions of convective fluid stability will in this case be determined by the conditions of free surface stability in the magnetic field.

In the linear approximation, surface-convective instability is described by equations (4.10), (4.12). For plane layers (cf. Fig. 5.5), boundary conditions (5.12) through (5.14) must include temperature dependences of the quantities. Let us consider that in a state of equilibrium ($v_0 = 0$), the fluid has a constant temperature gradient γ whose positive value corresponds to its direction from the free surface inside the fluid. Then the terms of the form appear in the rhs of equations (5.12). Appropriate differentiation and summation with regard to the continuity equation from (5.12) give one equation for $\partial\alpha/\partial x = (\partial\alpha/\partial T)\partial(T' - \gamma\xi)/\partial x$

$$\{\eta(\Delta_1 v_z - \partial^2 v_z/\partial z^2)\} = -\beta_\alpha \Delta_1(T' - \gamma\xi) \tag{5.41}$$

Pressure disturbances are found from the Navier-Stokes equations by applying the continuity equation

$$\Delta_1 P' = \Delta P' - \frac{\partial^2 P'}{\partial z^2} = \rho\frac{\partial}{\partial t}\frac{\partial v_z}{\partial z} - \eta\Delta\frac{\partial v_z}{\partial z} + \mu_0 M_0 \Delta_1 H' \tag{5.42}$$

and are substituted into the differentiated boundary condition of the balance of normal stresses (5.13) derived with allowance for temperature disturbances.

At $z = 0$, thermal boundary conditions in the linear approximation have the form

$$\{T' - \gamma\xi\} = 0; \{-\lambda\partial T'/\partial z\} = 0$$

in a conjugate statement, and for heat transfer by the Newton law

$$-\lambda\partial T'/\partial z = \alpha_T(T' - \gamma\xi)$$

Of greatest interest here is the magnetic field intensity component normal to the surface as it is responsible for surface instability. Therefore, assume

$$\vec{H}_0 = [0, 0, H_0], \vec{e}_H = [0, 0, 1]; \vec{M}_0 = [0, 0, M_0]$$

Make the equations dimensionless with allowance for the fact that, in this case, magnetic field perturbations are due to two reasons, namely disturbances of temperature (γl) and of free surface (M^*). Having in mind that surface disturbances, as a rule, lead to more essential field distortions, we take the quantity M^*l as a scale for the potential ψ. The velocity, temperature and time scales are κH, γl and l^2/κ.

For fluid layers unbounded in x and y directions, stability is considered relative to normal disturbances proportional to $\exp i\vec{\kappa}\vec{r}$ where $\vec{\kappa} = [\kappa_x, \kappa_y, 0]$. First, assume that above the MF surface there is a nonmagnetic gas with thermomechanic processes that may be ignored and the heat transfer obeys the Newton law. Consider also that for the potential of magnetic field disturbances in gas ψ_e the relations $\partial\psi/\partial_z = \kappa\psi_e$ are satisfied in case of normal disturbances. This enables us to eliminate this potential from the boundary conditions for the field in a fluid.

The dimensionless set of equations and the surface boundary conditions describing surface-convective stability of a horizontal plane-parallel layer is reduced to the form

$$\left.\begin{aligned} &\frac{1}{\mathrm{Pr}}\left(\frac{\partial^2}{\partial z^2} - \kappa^2\right)\frac{\partial v_z}{\partial t} = \left(\frac{\partial^2}{\partial z^2} - \kappa^2\right)^2 v_z - (\mathrm{Ra} + \mathrm{Ra}_m)\kappa^2\vartheta + \\ &+ \left(\frac{D_p}{L}\right)\kappa^2\frac{\partial\psi}{dz}; \\ &\partial\vartheta/\partial t - v_z = (\partial^2/\partial z^2 - \kappa^2)\vartheta; \; \partial^2\psi/\partial z^2 - A\kappa^2\psi - (L/\mu_r)\partial\vartheta/\partial z = 0 \end{aligned}\right\} \tag{5.43}$$

On a free surface $z = 0$

$$\left.\begin{aligned} & C_G\left(\frac{1}{\text{Pr}}\frac{\partial}{\partial t}-\frac{\partial^2}{\partial z^2}+3\kappa^2\right)\frac{\partial v_z}{\partial z}=-\left(1+\frac{\kappa^2}{\text{Bo}}\right)\kappa^2\zeta+ \\ & +\frac{\text{Si}}{\sqrt{\text{Bo}}}\mu_r\kappa^2\frac{\partial\psi}{\partial z}-\frac{\text{Si}}{\sqrt{\text{Bo}}}L\kappa^2\vartheta; \\ & \partial\zeta/\partial t=v_z;\quad(\partial^2/\partial z^2+\kappa^2)\,v_z=-\text{Ma}\kappa^2(\vartheta-\zeta); \\ & \partial\vartheta/\partial z=-\text{Bi}(\vartheta-\zeta);\ \mu_r(\partial\psi/\partial z)+\kappa\psi-L\vartheta-\kappa\zeta=0 \end{aligned}\right\} \quad (5.44)$$

The known dimensionless parameters are found as before; the Biot number $\text{Bi} = \alpha_T l/\lambda$ describes heat transfer on the free surface; $D_p = \mu_0\gamma^2 l^4 k^2/(\eta\kappa)$; the parameter $L = K\gamma l/M^*$ characterizes the relative contribution of temperature and surface magnetic field disturbances; $C_G = \eta\kappa(\rho g l^3)$ is the graviational number; $M^* = M_0(z = 0)$.

We have already introduced the earlier or so far known dimensionless numbers, yet not all of them are independent. Among them, in particular, is $D_p C_G\sqrt{\text{Bo}} = L^2\text{Si}$.

Below are the ranges of numerical values of the above numbers which may hold for MFs:

$$\begin{aligned} & \text{Pr}\simeq 10;\ \text{Ra}\simeq 0\div 10^6;\ \text{Ra}_m\simeq 0\div 10^6;\ D_p\simeq 0\div 10^3; \\ & \text{Ma}\simeq 0\div 10^3;\ \text{Bi}\simeq 0\div 10^3;\ C_G\simeq 10^{-7}\div 1; \\ & \text{Bo}\simeq 0\div 10^4;\ \text{Si}\simeq 0\div 10^3 \end{aligned}$$

High values of the gravitational number ($C_G \sim 1$) correspond to the layers of small thickness (less than 1 mm) in which thermocapillary processes play a significant role.

In the external uniform magnetic field, with the considered directions of vectors, a vertical temperature gradient of magnetic field intensity $G = K\gamma/\mu_r$ is formed in the fluid. In such a situation $\text{Ra}_m = D_p/\mu_r$.

If the lower layer boundary is solid nonmagnetic and isothermal, then at $z = -1$ the following boundary conditions must hold on it:

$$v_z = \partial v_z/\partial z = 0;\ \vartheta = 0,\ \mu_r\partial\psi/\partial z - k\psi = 0 \quad (5.45)$$

Equations (5.42)–(5.44) are solved as hyperbolic sines and cosines.

The most simple situation is come across when we can ignore the temperature dependences of density and magnetization ($K = \beta_\rho = 0$, $\text{Ra} = \text{Ra}_m = D_p = L = 0$) and concentrate on the thermocapillary and magnetic surface instability mechanisms. For this case, the equation of the neutral fluid stability curve, resolved relative to the Marangoni number at $A = 1$, has the form

$$\mathrm{Ma} = \frac{8\kappa(\mathrm{Bi}\ \mathrm{sh}\ \kappa + \kappa\ \mathrm{ch}\ \kappa)\ (\mathrm{sh}\ \kappa\ \mathrm{ch}\ \kappa - \kappa)R_\kappa}{8\kappa^4 C_G \sqrt{\mathrm{Bo}}\ \mathrm{ch}\ \kappa + (\mathrm{sh}^3\kappa - \kappa^3\ \mathrm{ch}\ \kappa)R_\kappa} \ * \tag{5.46}$$

where

$$R_\kappa = \sqrt{\mathrm{Bo}}/\kappa + \kappa/\sqrt{\mathrm{Bo}} - \mathrm{Si}(\mu\,\mathrm{th}\,\kappa + 1)/(2 + \frac{\mu^2+1}{\mu}\mathrm{th}\,\kappa)$$

In limited cases Si = 0 and Ma = 0, this relation, correspondingly, implies the equation of neutral curves for thermocapillary stability of a normal liquid and the above equation derived for the surface instability of an isothermal MF.

Minimization of rhs of (5.45) with respect to k specifies the critical values of the problem parameters. The main relationship Ma_* (Si) or Si_* (Ma) determines the interdependence of thermocapillary and surface instabilities. At small wave numbers ($\kappa \ll 1$) made dimensionless over the layer thickness, i.e., for a thin liquid layer, expression (5.46) gives

$$\mathrm{Ma} = \frac{2}{3}\ \frac{\mathrm{Bi}+1}{C_G \sqrt{\mathrm{Bo}}}\ \kappa\left[\frac{\kappa}{\sqrt{\mathrm{Bo}}} + \frac{\sqrt{\mathrm{Bo}}}{\kappa} - \frac{\mathrm{Si}}{2}\right] \tag{5.47}$$

whence follows the relationship between the critical values of the parameters

$$\mathrm{Ma}_* = \frac{2}{3}\ \frac{\mathrm{Bi}+1}{C_G}\left(1 - \frac{\mathrm{Si}^2}{16}\right);\ \kappa_* = \mathrm{Si}\sqrt{\mathrm{Bo}}/4 \tag{5.48}$$

As before, the thin layer corresponds to $\mathrm{Bo} \ll 1$. From (5.48) it follows that, with variations of the surface instability number Si from 0 to 4, the critical values of the Marangoni number deminish by the square law from the highest value $\mathrm{Ma}_* = (2.3) \times (1 + \mathrm{Bi})/C_G$, corresponding to normal liquid or no-field conditions, to zero.

Similarly, by solving equation (5.48) for Si we obtain the expression

$$\mathrm{Si}_* = 4\sqrt{1 - \frac{3}{2}\ \frac{\mathrm{Ma}\,C_G}{1+\mathrm{Bi}}} \tag{5.49}$$

to describe the destabilizing effect of the thermocapillary mechanism on surface stability of the fluid heated from below ($\gamma > 0$, $\mathrm{Ma} > 0$). With growing Ma, the critical values of Si are decreased from 4 to 0. In case of heating from above ($\gamma < 0$, $\mathrm{Ma} < 0$), the thermocapillary mechanism increases surface stability in the magnetic field ($\mathrm{Si}_* > 4$).

The relationships so derived point to strong independence of the thermocapillary and magnetic mechanisms of layer instability. As convective heat transfer from the free surface increases ($\mathrm{Bi} \to \infty$), the effect of the thermocapillary mechanism decreases.

* See Nomenclature p. 211 for explanation of notation for trigonometric functions.

The consideration of the thermogravitational convection mechanism yields the following critical relationships at small values of K:

$$\mathrm{Si}_* = 4\sqrt{1 - \frac{3}{2}\frac{\mathrm{Ma}\,C_G}{1+\mathrm{Bi}} - \frac{11}{40}\frac{\mathrm{Ra}\cdot C_G\,\mathrm{Bi}}{1+\mathrm{Bi}}}\;;\quad \kappa_* = \sqrt{\mathrm{Bo}}\;\mathrm{Si}/4 \tag{5.50}$$

They imply that, with heating from below (Ra > 0), the thermogravitational mechanism is an additional destabilizing factor, its effect being enhanced with the growing Biot number.

In quantitative respect, the degree of the effect of the thermocapillary and gravitational convection mechanisms is greatly determined by the gravitational number C_G which characterizes deformability of the free surface by the convective motion. Large values of C_G correspond to extensive surface distortions. At $C_G = 0$, fluid motion does not deform the surface and, hence, does not affect magnetic instability ($\mathrm{Si}_* = 4$, $\kappa_* = \sqrt{\mathrm{Bo}}$). However, the free surface deformed by magnetic instability is sure to give rise to convective motion in a nonisothermal fluid. In this case, the convective instability threshold is determined by the surface instability threshold.

For a thick fluid layer ($\kappa \gg 1$) whose thickness is much greater than the capillary radius ($\mathrm{Bo} \gg 1$), the critical dependences accurate to the terms squared with respect to $\mathrm{Ra}C_G/\sqrt{\mathrm{Bo}}$ are of the form:

$$\mathrm{Si}_* = 2(1+\mu^{-1})\left[1 - \frac{3\,\mathrm{Ra}\,C_G\,\mathrm{Bi}}{8\sqrt{\mathrm{Bo}}(1+\mathrm{Bi})}\right];\quad \kappa_* = \sqrt{\mathrm{Bo}}\left[1 - \frac{3\mathrm{Ra}\,C_G\mathrm{Bi}}{4\sqrt{\mathrm{Bo}}(1+\mathrm{Bi})}\right] \tag{5.51}$$

In thick fluid layers, the effect of the thermocapillary mechanism disappears, while the thermogravitational mechanism destabilizes, as ever, the surface when the equilibrium temperature gradient is directed from the free surface inside the fluid (heating from below) and stabilizes it when the gradient direction is opposite.

Changing thermal conditions on the lower boundary (for instance, constant heat flux prescribed on it) mainly bring about qualitative differences in the behavior of the processes under study.

5.5 CAPILLARY DISINTEGRATION OF MAGNETIC FLUID JETS

As far as free fluid surface control by a magnetic field is concerned, of special importance are the cases of preventing instability and disintegration of fluid volumes under the action of capillary forces. By striving to impart to the fluid a spherical shape, as one having the smallest surface, the latter often cause this volume to disintegrate into separate droplets. A classical example of such a situation is the disintegration of a cylindrical fluid jet into droplets. This situation is of great practical importance and became a subject of profound research as far back as the last century.

The subject of our study is the stability of cylindrical MF layers in nonuniform and uniform magnetic fields as applied to the problems of controlling the stability of cylindrical jet and film flows. A nonuniform magnetic field affects the flow through the volume magnetic force induced in the fluid, while a uniform magnetic field hinders the development of free surface disturbances.

Worthy of special mention is the fact that immovable cylindrical layers with free surface may be created with the use of magnetic fluids in laboratory conditions and different capillary disintegration conditions may be simulated. Immobility of the fluid layer greatly facilitates the experiment. In a normal liquid, the free cylindrical surface is only formed as a result of an expendable jet or film flow. In laboratory experiments, it is possible to eliminate the gravitational force and make the cylindrical layer hydraulically weightless, i.e., put it into the fluid of the same density.

It is simple enough to create a cylindrical MF column around a live cylindrical conductor of radius R. The magnetic field of line current I has only an azimuthal component $H_\theta = I/(2\pi r)$ and intensity gradient $|\nabla H| = I/2\pi r^2$ directed to the conductor axis. Under the action of this gradient, the MF is distributed as a cylindrical layer of a radius a around the live conductor (Fig. 5.7). At a small radius of the live conductor ($R \ll a$), the situation is equivalent to a continuous fluid column, while at $(a - R) \ll R$ it is equivalent to a thin film. The results obtained for this geometry are applicable both to jet and film fluid flows.

Stability of such a layer is determined by the competing action of magnetic and capillary forces. Magnetic pressure is of the order of $\mu_0 M|\nabla H|\alpha$, the capillary one of α/α Their ratio $\mathrm{Bo}_m = \mu_0 M|\nabla H|\alpha^2/\alpha$ may be regarded as the magnetic bond number by analogy with the gravitational $\mathrm{Bo} = \rho g \alpha^2/\alpha$. At $\mathrm{Bo}_m > 1$, magnetic forces and, at $\mathrm{Bo}_m < 1$, capillary forces are predominant.

It is natural that theoretical consideration of the above problems should be carried out in the cylindrical coordinate system (r, θ, z), with the axis z being the conductor axis. In cylindrical geometry, the Maxwell equations have an exact solution of the form

$$\vec{H} = [H_r = A/r,\ H_\theta = I/(2\pi r),\ Hz = \mathrm{const}]$$

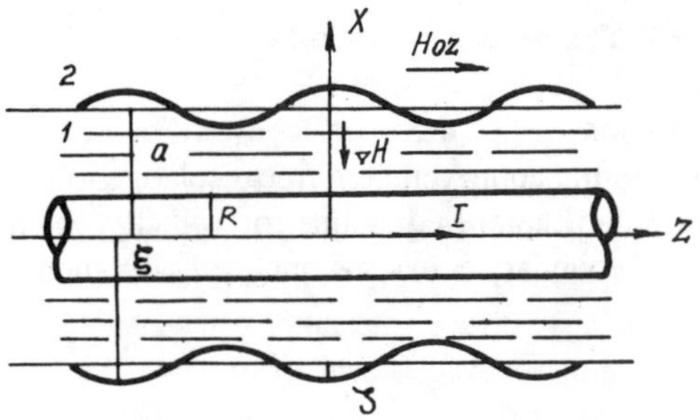

FIG. 5.7

If the surface equation is $r = \xi(z)$, then its curvature and the normal vector are specified as

$$R_1^{-1} + R_2^{-1} = (1 + \xi'^2)^{-3/2}[(1 + \xi'^2)/\xi - \xi''];$$
$$\vec{n} = \left[\frac{1}{\sqrt{1+\xi'^2}}, 0, \frac{-\xi'}{\sqrt{1+\xi'^2}}\right] \tag{5.52}$$

where the prime denotes differentiation over z.

While studying layer stability in the linear approximation relative to wave perturbations, let us emphasize that the general solution to the Laplace equation $\Delta\varphi = 0$, for perturbations $\varphi(t, r, \theta) = f(r) \exp i(\omega t - m\theta)$ that are periodic with respect to the angle θ, is

$$f(r) = C_1 r^m + C_2 r^{-m}$$

and for axisymmetric perturbations $\varphi(t,r,z) = f(r) \exp i(kz - \omega t)$ it is expressed through modification of Bessel's function

$$f(r) = C_1 I_0(\kappa r) + C_2 K_0(\kappa r)$$

Bessel's function K_0 gives solutions attenuating at infinity; I_0 yields solutions limited at $r = 0$.

The methods of studying the stability of mechanical equilibrium of a cylindrical MF column is the same as in the case of plane layers.

Surface Stabilization with a Nonuniform Magnetic Field

The field of line current $\vec{H} = [0, I/2\pi r, 0]$ is everywhere tangential to the free fluid surface and in case the axial symmetry of the problem is retained, there are neither magnetic field disturbances, magnetic pressure jump nor magnetic mechanism of surface instability to be observed. The magnetic field effect only manifests itself in the form of volume magnetic force. The characteristic field gradient $G = I/(2\pi a^2)$ and the magnetic Bond number $\mathrm{Bo}_m = \mu_0 M^* I/(2\pi\alpha)$, while for the linear magnetization law $M = \chi H$ we have

$$M^* = \chi H(r = a); \ \mathrm{Bo}_m = \mu_0 \chi I^2/(4\pi^2 \alpha a)$$

Should we introduce small deviations, ξ, of the interface from the equilibrium value $\xi = \xi - a$, the problem in the approximation linear relative to perturbation amplitudes is described by the Laplace equations for velocity potentials $\Delta\varphi = 0$ and the following set of boundary equations obtained with the use of the Cauchy-Lagrange integral:

at $r = R$

$$\partial\varphi_1/\partial r = 0$$

at $r = a$

$$\rho_1(\partial\varphi_1/\partial t) - \rho_2(\partial\varphi_2/\partial t) - \mu_0\chi I^2\zeta/(4\pi^2 a^3) + \alpha(\zeta/a^2 + \zeta'') = 0;$$

$$\partial\zeta/\partial t = \partial\varphi_1/\partial r = \partial\varphi_2/\partial r \tag{5.53}$$

Subscripts 1 and 2 correspondingly refer to the MF and the liquid above it.

Substitution into (5.53) of the solutions to the Laplace equations for axisymmetric wave disturbances, taking account of the required attenuation φ at infinity, results in the following dispersion equation written in the dimensionless form as:

$$\Omega^2 = \left[\frac{I_0(s)K_1(\sigma_R s) + I_1(\sigma_R s)K_0(s)}{I_1(s)K_1(\sigma_R s) - I_1(\sigma_R s)K_1(s)} + \right.$$

$$\left. + \sigma_\rho \frac{K_0(s)}{K_1(s)}\right]^{-1} s(\mathrm{Bo}_m - 1 + s^2) \tag{5.54}$$

The dimensionless frequency Ω and the wave number s are related to denominate quantities by relations $\Omega^2 = \rho_1 a^3\omega^2/\alpha$ and $s = ka$. Besides, $\sigma_R = R/a$, $\sigma_\rho = \rho_2/\rho_1$.

The sign in the rhs of (5.54) is only determined by the relation $\mathrm{Bo}_m - 1 + s^2$ as the expression before it is always positive. At $\mathrm{Bo}_m > 1$, this relation is positive. Ω is material and the MF column is stable. At $\mathrm{Bo}_m < 1$, there will always be a wave number for which $\mathrm{Bo}_m - 1 + s^2 < 0$, i.e., Ω becomes imaginary and stipulates a monotonic growth of perturbations making the fluid column unstable and causing it to disintegrate into droplets. From this viewpoint, the critical magnetic Bond number is unity. With decreasing Bo_m number, the column becomes unstable for all disturbances whose dimensionless wavelength $\Lambda = 2\pi/s$ is in excess of $2\pi/(1 - \mathrm{Bo}_m)^{1/2}$.

Most detrimental are the disturbances with the shortest time of development, i.e., disturbances with the wavelength to which corresponds the negative value of Ω^2 maximum in magnitude.

From these standpoints, expression (5.54) can easily be analyzes in a limited case of a thin cylindrical layer $1 - \sigma_R = (a - R)/R \equiv \delta \ll 1$. in a nonmagnetic gas ($\sigma_\rho = 0$). Accurate to the terms of the order of δ^2 we set from (5.54)

$$\Omega^2 = \delta s^2(\mathrm{Bo}_m - 1 + s^2) \tag{5.55}$$

Hence, the characteristics of most detrimental disturbances with subscript $_*$ are estimated as:

$$\Lambda_* = 2\pi\sqrt{2}/\sqrt{1 - \text{Bo}_m};\ |\Omega_*| = \sqrt{\delta}\,(1 - \text{Bo}_m)/2 \qquad (5.56)$$

At $\text{Bo}_m = 0$, which corresponds to a nonmagnetic fluid or no-field condition, $\Lambda_* = 2\pi\sqrt{2} \approx 8.9$. This practically coincides with the known value $\Lambda_* = 9.01$ for a continuous cylindrical column of mormal fluid. Hence, we may conclude that the layer thickness does not in fact influence the length of perturbations leading to its disintegration. As the Bo_m number approaches unity, Λ_* increases tending to infinity. For a fluid column of finite length this means a few number of droplets it disintegrates into as a result of instability. The characteristic time it takes perturbations to develop $|\Omega_*|^{-1}$ grows tending to infinity as Bo_m increases. A thinner layer thickness δ also hinders the disintegration process.

The instability characteristics of a cylindrical MF layer of finite thickness, calculated by formulae (5.54), and the experimental results are presented in Figs. 5.8 and 5.9. The magnetic field can change the size of droplets of the disintegrated layer within a wide range.

Uniform Axial Field H_z Effect on Layer Disintegration

A uniform magnetic field directed along the layer must exert a stabilizing effect on the latter by hindering the development of surface disturbances periodic along their direction, i.e., of those disturbances which cause the layer to disintegrate into droplets. The effect of such a field is described by the surface stabilization number which for the case may be formulated as

$$\text{St} = \mu_0 M_z^2 (a - R)/\alpha$$

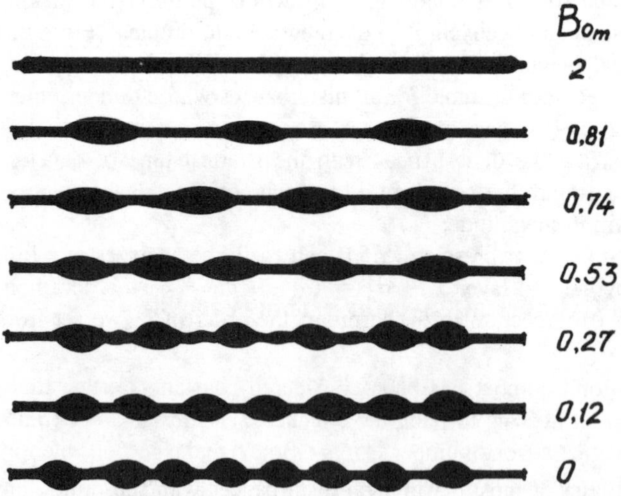

FIG. 5.8

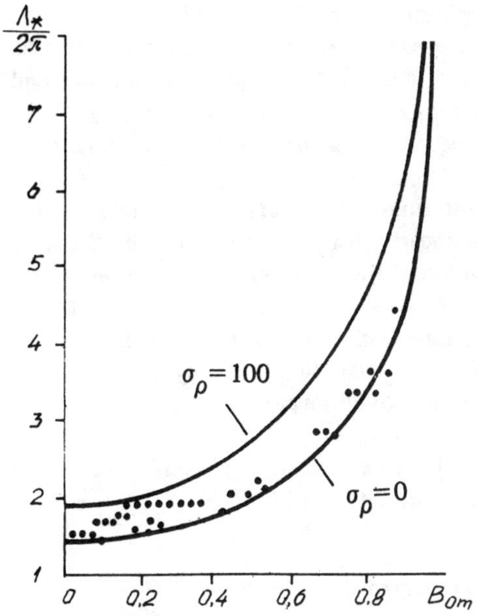

FIG. 5.9

The solution of the stability problem with regard to magnetic field perturbations yields, for example, the following dispersion equation for a thin layer

$$\Omega^2 = \delta s^2[\mathrm{Bo}_m - 1 + (1 + \mathrm{St}/\mu)s^2] \tag{5.57}$$

which gives the following wave characteristics of most detrimental disturbances

$$\Lambda_* = 2\pi\sqrt{2}\sqrt{\frac{1 + \mathrm{St}/\mu}{1 - \mathrm{Bo}_m}}\ ; \quad \Omega_* = \sqrt{\delta}\,\frac{1 - \mathrm{Bo}_m}{\sqrt{1 + \mathrm{St}/\mu}} \tag{5.58}$$

The comparison of the expressions obtained with (5.56) shows that the stabilizing effect of the axial magnetic field manifests itself in a longer wavelength and longer time it takes the perturbations to develop and cause the layer to disintegrate into droplets.

Disintegration of a Cylindrical Layer in a Time-Variable Magnetic Field

So far the magnetic Bond number has been considered constant. Certain time modulation of the current flowing through the conductor ensures an appropriate law for varying the magnetic Bond number $\mathrm{Bo}_m = \mathrm{Bo}_m(t)$ and is responsible for the new specific features of the layer disintegration process.

In case of periodic temporal variations of the current, the equations describing the layer stability take on the form of the Mathieu equation. One of the main properties of its solution is the possibility of describing the parameteric resonance in the system. Then, even at average values of the magnetic Bond number greater than unity, parametric instability may develop in it.

High-frequency modulation of the current in the conductor stabilizes the layer.

If the magnetic field changes with time monotonely, for example, so that the magnetic Bond number vanishes linearly from its threshold value $Bo_{m*} = 1$, then the layer disintegrates in a different way. By prescribing the rate of Bo variation we can, in fact, set the characteristic time of the process. As a result, the disturbances, whose characteristic time is close to the time of Bo_m variation, will develop. With decreasing rate of Bo_m variation, i.e., by increasing the characteristic time prescribed, the wavelength of developing disturbances grows (cf. Fig. 5.10).

It follows from the above that the magnetic field ensures an effective control of capillary disintegration of magnetic fluid jets and films.

5.6 DESTRUCTION OF EQUILIBRIUM MAGNETIC FLUID FORMS IN A UNIFORM MAGNETIC FIELD

The essence of the phenomenon is the following. When no magnetic field exists, a limited volume of magnetic capillary fluid placed in a vertical gravitational field on a solid surface takes on a stable equilibrium form. When a vertical

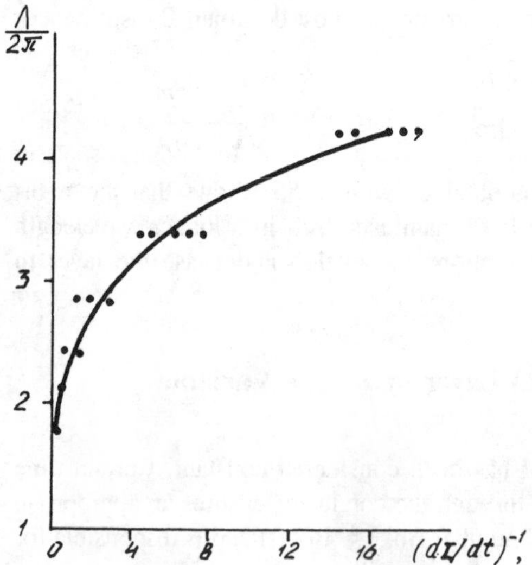

FIG. 5.10

uniform external magnetic field is applied and as its intensity increases, the MF volume tends to extend along the direction of the magnetic field. For all MF volumes in the absence of a magnetic field, there exists a single-bonded equilibrium fluid form. However, when the magnetic field intensity is in excess of a certain critical value dependent on the MF properties and its volume, the single-bonded form becomes topologically unstable: a vertical groove is formed on one side of the volume, beginning from the point of contact with the solid surface; along the groove, expanding with time, the droplet breaks down into two droplets approximately equal in volume which are stationary at some distance from each other. With the droplet volume increasing, the critical value of the magnetic field intensity, which is necessary for the droplet to break down, decreases. There exists a critical droplet volume: droplets of smaller volumes do not break down at infinitely large values of magnetic field intensity.

The equilibrium form breaks down at any rate of magnetic field increase, small as it may be, in a very short range of magnetic field intensities. The breakdown behavior allows a conclusion that the single-bounded MF volume disintegrates as a result of a specific topological instability of a steady equilibrium state.

The breakdown of an equilibrium form was experimentally observed on various MF droplets when intensity of the uniform vertical magnetic field was varied within $10 < H_0 < 500$ kA/m. At H_0 = const, the observation time interval was chosen to be different for different fluids and was 15 min. for low-viscosity MFs and 30 min. for high-viscosity MFs. At a critical value H_0, the time of droplet disintegration was essentially dependent on the fluid properties and, namely, on the MF viscosity and the wetting angle. While for experiments with a low-viscosity MF, which does not wet the substrate, the time of disintegration was several minutes, then for a high-viscosity MF having good substrate wetting properties it was several hours. The break of a single 0.35 cm^3 droplet in a vertical uniform magnetic field is illustrated by photographs in Fig. 5.11. Each photo shows two mutually perpendicular projections obtained with a system of mirrors. Fig. 5.11a shows the droplet at H = 8 kA/m magnetic field intensity, i.e., much less than the critical value. The next photos show the droplet at a time interval of about 10 s. At constant mgnetic field intensity H = 85 kA/m, we can see typical forms of the droplet it took on in the course of its breakdown.

FIG. 5.11

The formation, motion and disintegration of a magnetic fluid droplet are described by the set of equations (2.34) and (2.49) subjected to conditions (2.41). It was used to study the form of droplets and their breakdown in the following approximation. The droplet form was found from (2.49) on the assumption that the magnetic field intensity inside the MF is equal to the external field intensity H_0. Then, the magnetic field intensity in the droplet of the known form was estimated from (2.34) and (2.41). In the magnetic field so defined, the extreme properties of the potential (free) energy of the system, assumed equal to the potential energy of the MF, were studied:

$$U = \sigma_{ij} S_{ij} + \int_{\Omega} (\Pi_g + \Pi_m) \, d\Omega$$

where σ_{ij} are the surface tension coefficients at the interface of the ith and jth continuous media; S_{ij} their contact area; Π_g, Π_m the specific (volume) densities of the potential MF energy due to the gravitational and magnetic fields, respectively,

$$\Pi_g = \rho g z \; ; \; \Pi_m = -\mu_0 \int_0^H M(H) \, dH$$

with the assumed Langevin law

$$M(H) = M_s [\mathrm{cth}(H/H_s) - H_s/H]$$

The mathematical model thus built allows the following conclusion to be drawn.

As the intensity of an external vertical magnetic field grows, the MF droplet on a horizontal surface elongates upward yet remains stationary. Nonuniformity of the magnetic field inside the droplet grows and gives rise to a volume force $\mu_0 (M\nabla)H$. It is localized in two regions: at the very top of the droplet it is directed upward and causes the upper droplet nose to elongate still more, while at the droplet base near the solid surface it is horizontal and radial from the center. This force $\mu_0 (M\nabla)H$, which attempts to break the droplet into different directions, may be compared to the centrifugal force during axial rotation of the droplet.

When the magnetic field intensity achieves its critical value, this narrowly localized force assumes such great values that it cannot be steadily balanced by the capillary force. Calculation (3.6) for this case shows the onset of instability at those magnetic field intensities which qualitatively correspond to the experimental results.

The critical magnetic field intensity grows with the decreasing droplet volume. Droplets of rather small volumes do not disintegrate, no matter how large magnetic field intensities may be. Better wettability of the solid surface with the droplet volume constant, corresponds to larger critical values of the magnetic field intensity.

5.7 RESONANCE PHENOMENA

As long as periodic time-dependent magnetic fields can be established, forced vibrational motions may be induced in a magnetic fluid. In many situations, the MF is a system possessing natural vibrations. All this provides prerequisites for a number of resonance phenomena to take place in the fluid. Some of them will be considered below.

Vibrations and Resonance Droplet Disintegration

A uniform magnetic field causes a spherical MF droplet to elongate along the field lines. The elongation of the droplet is greater the higher the field intensity. Periodic extension and compression of the droplet will be observed in a periodic time-independent magnetic field. The droplet is compressed, i.e., returned to the initial state with the field off, under the action of capillary forces. With such forced vibrations, the magnetic field energy transforms into the kinetic energy of droplet motion.

On the other hand, any fluid volume is characterized by a discrete frequency spectrum ω of natural capillary vibrations of the free surface. This spectrum is determined by the surface form. For example, for a spherical fluid volume of a radius a it is expressed by the relation [2]

$$\omega^2 = (\alpha/\rho a^3)n(n - 1)(n + 2) \quad \mathrm{n} = 1, 2, 3 \ldots$$

The coincidence of the frequency of the acting force, in this case of a magnetic field, with one of the natural frequencies of surface vibrations, brings about an instant of resonance. A maximum of the field energy transforms into kinetic energy of the fluid volume in the resonance region, which leads to a drastic increase in the vibration amplitude. This energy is sufficient for a droplet to break down into parts. The number of droplet parts m is determined by the frequency of the field which excites the corresponding harmonic of surface waves. This number grows with frequency of field variations.

This process was experimentally studied under the conditions of hydraulic imponderability using high-speed photography. Fig. 5.12 shows the cinegrams of droplet vibrations that cause its disintegration (within a period of single vibration). For a droplet 1–3 mm in radius, resonances were registered in the frequency range between 1 and 15 Hz consistent with natural capillary vibrations. The spectrum of experimental dimensionless resonance frequences $\Omega = \omega\sqrt{\rho a^3/\alpha}$ is presented in Fig. 5.13. As the vibration frequency increased, resonance could be observed on higher harmonics and was accompanied by the breakdown of the drop into a greater number of parts m. The greatest number of secondary droplets registered was seven.

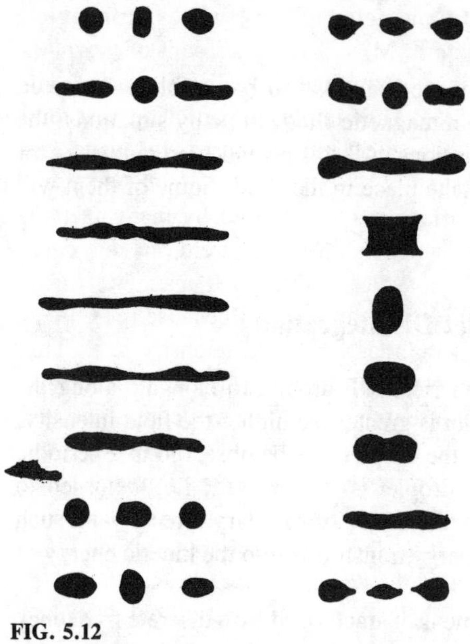

FIG. 5.12

Resonance Excitation of Surface Waves

Waves on the MF surface were considered in 5.2–5.5 in connection with its stability. If the values of parameters are less than the critical ones, then the above dispersion equations describe a steady propagation of surface waves under appropriate conditions. Thus, free gravitational-capillary waves described by the dispersion equation $\omega^2 = \kappa g + \alpha\kappa^3/\rho$ propagate on the surface of an infinitely

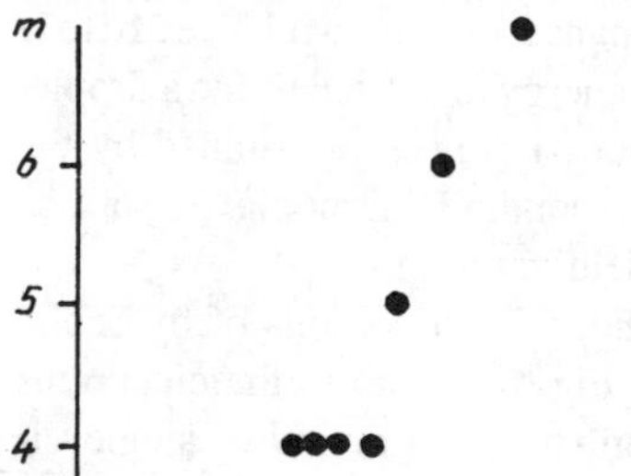

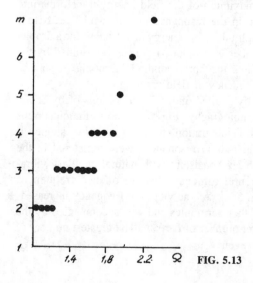

FIG. 5.13

thick fluid layer. On the other hand, magnetocapillary waves, for which $\omega^2 = [\kappa\mu_0 M|\nabla H| + \alpha\kappa^3]/\rho$, propagate in MFs. Taking account of magnetic field disturbances, the dispersion equations for surface waves have a more complex form (cf., for example, formulae (5.5). They imply, in particular, that for a prescribed wavelength the tangential magnetic field increases the wave frequency, while the normal one decreases it.

Generation of waves on the MF surface may be caused by many factors, one of them being a time- and space-dependent magnetic field. In this case, resonance may occur under the action of a variable applied magnetic force. Physically, the forced waves on the MF surface are excited because pressure at any point in the fluid is proportional to the magnetic field intensity in it. This means that the free fluid surface, as a constant-pressure surface in a space-variable magnetic field, will be wavy, representing a system of depressions and elevations that correspond to minima and maxima of the field intensity. If the magnetic field is a running wave, i.e., represents a system of minima and maxima moving along the surface, it will entrain the system of depressions and elevations on the fluid surface thus exciting the forced running surface wave with the wave characteristics determined by the wave characteristics of the field. A forced standing is formed on the fluid surface in the magnetic field described by the standing wave.

On the other hand, the free fluid surface is a system having natural vibrations (gravitational-capillary, magnetocapillary). When the characteristics of the forced surface wave, excited by the running magnetic field, coincide with the characteristics of the proper waves, the conditions are most favorable for energy transfer from the magnetic field to the wave and its excitation will be resonant in nature. The condition for the onset of wave resonance implies a relationship between the frequency and the wavelength of the acting force that coincides with the dispersion equation for proper waves in the system. This, in principle, discriminates the system from resonance in localized systems (for instance, pendulum) when there is but a single vibrational characteristic of a force, i.e., frequency. In the case under consideration, the resonance conditions are achieved by varying both the vibrational frequency of the acting force and its wavelength, i.e., both time and spatial characteristics.

The magnetic field as a running or standing wave is described by the Laplace equation solved for the magnetic field potential $\Delta\psi = 0$:

$$\psi = (C_1 e^{\kappa z} + C_2 e^{-\kappa z})\cos(\kappa x - \omega t)\text{—running wave;}$$

$$\psi = (C_1 e^{\kappa z} + C_2 e^{-\kappa z})\cos\kappa x \cdot \cos\omega t\text{—standing wave.}$$

Such fields are induced by special inductor systems.

Let the magnetic field be of the following configuration

$$\left.\begin{aligned} &H_x = H^* + H_a e^{-\kappa z}\cos(\kappa x - \omega t);\; H_z = H_a e^{-\kappa z}\sin(\kappa x - \omega t);\\ &H^2 = H^{*2} + H_a^2 e^{-2\kappa z} + 2H^* H_a e^{-\kappa z}\cos(\kappa x - \omega t)\end{aligned}\right\}\quad (5.59)$$

In the situation under consideration, the magnetic field develops a ponderomotive force as it is nonuniform along the axis x. The effect of magnetic field disturbances due to surface curvature will be a quantity of the second order of smallness and is omitted here. Neither are considered the magnetic pressure jump and the capillary force.

Then, from the equations describing surface waves on a magnetic fluid, we obtain the following equation for the velocity potential in the linear approximation for a horizontal plane-parallel layer (Fig. 5.14):

$$\left[\hat{g} \frac{\partial \varphi}{\partial z} + \frac{\partial^2 \varphi}{\partial t^2} - \omega \frac{\mu_0 \chi H^* H_a}{\rho} e^{\kappa z} \sin(\kappa x - \omega t) \right]_{z=0} = 0 \qquad (5.60)$$

where

$$\hat{g} = g + \mu_0 \chi \kappa H_a^2 / \rho$$

The addition to gravitational acceleration is stipulated by nonuniformity of the running magnetic field normal to the surface.

The solution of the Laplace equation for the velocity potential $\Delta\psi = 0$ of an infinitely thick layer, which satisfies boundary condition (5.58), is of the form

$$\varphi(x, z, t) = \varphi_a e^{\kappa z} \sin(\kappa x - \omega t) \qquad (5.61)$$

where

$$\varphi_a = -\omega \frac{\mu_0 \chi H^* H_a / \rho}{\omega^2 - \kappa \hat{g}}$$

It follows from this relation that at the running magnetic field frequency and wavelength satisfying the dispersion equation of surface gravitational waves $\omega^2 = \kappa \hat{g}$, the aplitude, φ_a, of excited waves goes into infinity, thus testifying to the onset of resonance in the system.

The amplitude of the fluid surface waves is determined from the kinematic condition

$$\zeta = \int \frac{\partial \varphi}{\partial z} dt = -\frac{\kappa}{\omega} \varphi_a \cos(\kappa x - \omega t) = -A \cos(\kappa x - \omega t)$$

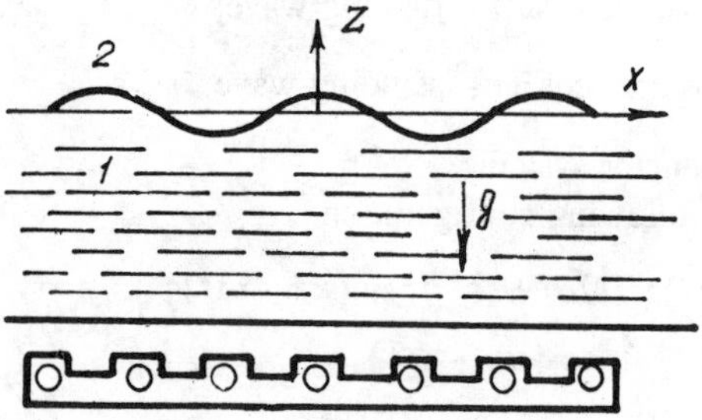

FIG. 5.14

For a fluid layer of a finite thickness l, limited by a solid surface from below and by nonmagnetic fluid bulk 2 from above, the resonance condition is of the form

$$\omega^2 = \frac{(\rho_1 - \rho_2)\kappa g + \alpha\kappa^3}{\rho_1 + \rho_2 \operatorname{th}\kappa l} \operatorname{th}\kappa l \qquad (5.62)$$

The experimental resonance curves so obtained are plotted in Fig. 5.15. The excited wavelength was 30 mm. The magnetic field as a standing wave was set up by a linear system of electromagnets. The distance between the magnets was controlled and the wavelength was fixed. A sharp increase of the surface vibration amplitude A was achieved at a frequency of 10 Hz.

Resonance frequency vs wavelength, calculated by formula (5.60), is presented by a solid line in Fig. 5.16. In the same plot, the dots stand for the experimental results which agree well with theoretical data.

Taking account of viscosity in theoretical consideration provides a limited amplitude of surface vibrations in the resonance

$$\xi_{a.res} = \mu_0 \chi H^* H_a / (4\rho\sqrt{\hat{g}}\, \nu\kappa^{3/2})$$

It is worth noting in conclusion that wave generation on the MF surface with magnetic fields of various configurations is a good means of modelling wave processes, including low and high tides caused by gravitational forces as well as of different accelerations in a normal fluid.

Resonance Sound Excitation

Periodic pressure variations caused by a variable nonuniform magnetic field in a fluid being compressed, will generate acoustic vibrations. If the magnetic field

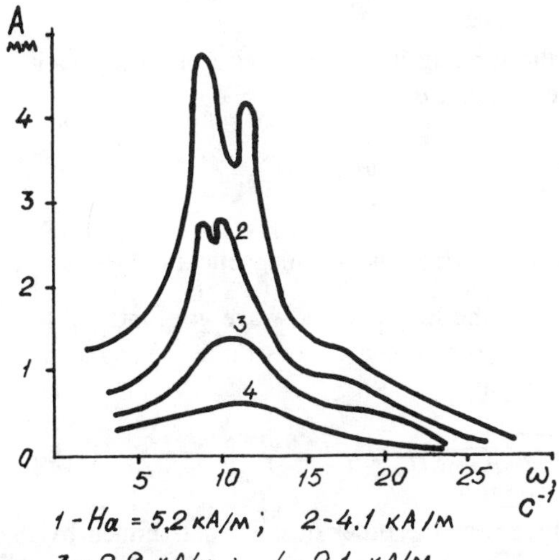

FIG. 5.15

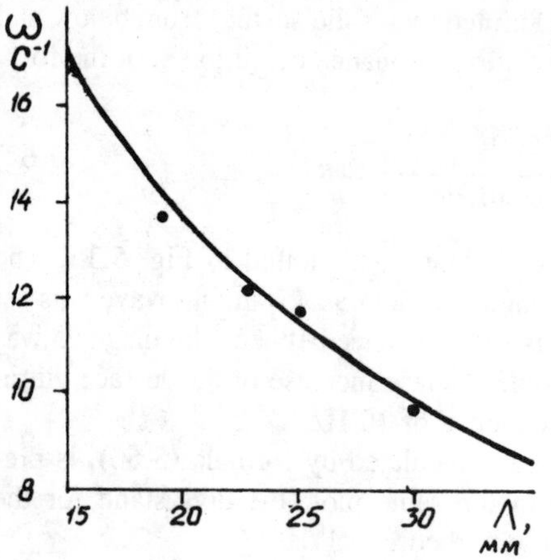

FIG. 5.16

has the form of a running or a standing wave, then a forced running or a standing sound wave is generated in an infinite fluid with wavelength- and frequency-dependent characteristics. When these characteristics satisfy the dispersion equation of natural acoustic vibrations, the excitation of these waves will become resonant. In practice, this will take place when the phase velocity of the field is equal to the sound velocity in fluid

$$c_{sw} = \sqrt{\partial P/\partial \rho}$$

In the presence of a volume magnetic force, the one-dimensional linear acoustic wave equation for an ideal fluid will have the form

$$\frac{\partial^2 v_x}{\partial t^2} - c_{sw}^2 \frac{\partial^2 v_x}{\partial x^2} = \frac{\mu_0 \chi}{2\rho} \frac{\partial^2 H^2}{\partial x \partial t} \tag{5.63}$$

If the field is described by the running wave $H^2 = H^{*2} + H_a^2 \cos(\kappa x - \omega t)$, then the forced sound wave is found from the expression

$$v_x = v_a \cos(\kappa x - \omega t),$$

where $v_a = \dfrac{\mu_0 \chi H_a^2}{2\rho} \dfrac{u}{c_{sw}^2 - u^2}$; $u = \omega/k$ – is the magnetic field phase velocity. At $u = c_{sw}$, the amplitude of the forced sound wave goes into infinity, which points to the resonance in the system. In a viscous fluid, the sound wave amplitude in the resonance is related as

$$v_a = A_H(\mu_0 \chi H_a^2/\kappa\eta) \tag{5.64}$$

where A_H is the numerical coefficient of the order of unity determined by the field configuration. If the field vibration amplitude is of the order of 10^4 A/m, then $v_a \sim 10^2$ m/s.

The intensity of the sound produced by the open end of a tube filled with fluid is

$$J = \rho_c S^2 \overline{(\partial v/\partial t)^2}/4\pi C_c \quad (5.65)$$

where S is the cross-sectional area of the tube; $\overline{(\partial v/\partial t)^2}$ the squared mean of the time derivative of velocity at the tube end; ρ_c, c_c, the density and velocity of sound in the medium into which the sound is radiated. With the sound radiated into the air at 10^4 Hz for $S = 1$ cm^2, $J \sim 1$ W.

Resonance Excitation of Sound in a Closed Vessel

Resonance excitation of sound in a magnetic fluid may also be effected when the latter is in a closed vessel. The vessel walls pose as a resonant cavity and determine the resonance conditions. The variable force applied may be other than periodic in space. It may be created, for example, by a magnetic field having constant amplitude of the intensity gradient $\partial H/\partial x = G \sin\omega t$. If the vessel is a parallelepiped with dimensions L, a and b, then the resonance frequency spectrum is specified by the expression

$$\omega^2 = c_{sw}(\kappa_l^2 + \kappa_m^2 + \kappa_n^2)$$

where

$$\kappa_l = (2l - 1)\pi/2L; \ \kappa_m = \pi m/a; \ \kappa_n = \pi n/b; \ l, m, n = 1, 2, 3$$

Resonance is achieved when the acting force frequency coincides with one of the natural frequences of the cavity. Of practical importance is the resonance frequency at which the amplitude of sound vibrations is maximum. Whether some or other resonance harmonic is maximum, depends on the spatial distribution of the acting force.

Field gradient amplitude, being constant, the amplitudes of physical quantities in the sound wave decrease with the growing harmonic number to achieve maximum value at $l = m = n = 1$.

This phenomenon has been studied experimentally. Round glass cylinders 48 to 84 mm dia were used as resonant cavities. A nonuniform variable magnetic field was set up by a solenoid. The field intensity amplitude was 2.4 kA/m and the current flowing through the solenoid was 10 A. The current frequency ranged between 16 and 26.7 kHz. The experiments have shown that in such a cylindrical resonant cavity with one free boundary, resonance excitation of ultrasonic vibrations can be effected in a ferromagnetic fluid. Sound pressure in

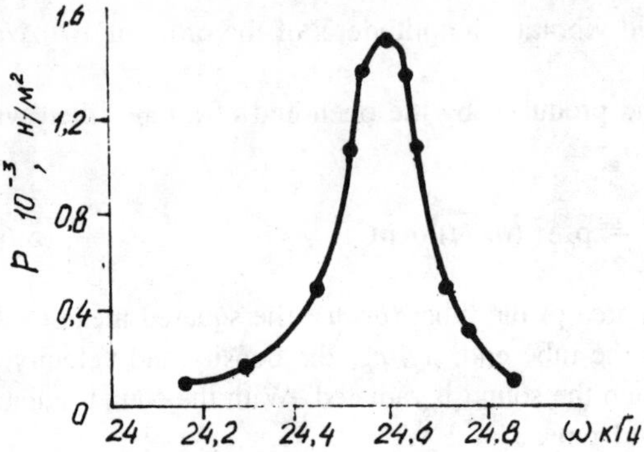

FIG. 5.17

the fluid drastically grows at resonance to 5×10^3 N/m^2, and energy density in the resonant cavity increases to 10^{-1} J/m^3. Fig. 5.17 shows the plotted experimental resonance curves which testify to high quality of the system.

The MF-based acoustic radiators display good performance irrespective of the environment; they do not exhibit stresses inherent in solid radiators and can replace the latter.

PART TWO

NONEQUILIBRIUM MAGNETIZATION

CHAPTER

SIX

EQUATIONS OF THERMOMECHANICS OF MAGNETIC FLUIDS WITH RELAXING MAGNETIZATION IN LOW-FREQUENCY APPROXIMATION

The phenomena considered in the previous sections are stipulated by their interaction with the field of equilibrium magnetization. However, the experiments reveal phenomena that can hardly be explained without taking account of magnetization relaxation when a magnetic fluid moves in magnetic fields or when affected by nonstationary fields. In this paper we shall develop a mathematical model allowing for magnetization relaxation, with minimal complication of ferrohydrodynamics equations as compared to the equilibrium magnetization approximation. It is a simplifying fact here that the characteristic hydrodynamic times as well as the times of magnetic field variation are, as a rule, in excess of the relaxation time of magnetization. In this case, an equation, linear in dynamic variables can be derived for magnetization. This equation is in good correlation with the constituent hydrodynamic equations and has a wide field of applications.

6.1 MICROSCOPIC CONCEPTS OF MAGNETIZATION RELAXATION

In this section, in order to introduce basic microscopic parameters for the dynamic component of MF magnetization, we shall discuss relaxation processes as applied to a single microscopic ferromagnetic particle suspended in a nonmagnetic viscous fluid. Detailed consideration to this problem is given in survey [14].

The volume of a particle is one of its most important characteristics. Three characteristic volumes should be distinguished for MF-forming particles. The ferromagnetic material volume V_f, which determines the magnetic properties of a particle, manifests itself during fluid magnetization measurements. The ferromagnetic core is coated with a layer of nonmagnetic oxide whose thickness is of the order of crystal lattice period (~0.5 nm). The oxide density is comparable with the density of the ferromagnetic material. Therefore, in measuring the MF density we can observe a particle solid matter volume V_s equal to the sum of volumes of the ferromagnetic core and the oxide shell. Long-chain surfactant molecules are adsorbed on the particle surface. These molecules form a nonmagnetic deformable layer of the order of the molecular length around the solid matter. This layer, whose thickness may be estimated by the oleic acid molecule length (~2.0 nm), prevents the particles from coalescing. The volume including the solid matter and the related polymer layer are observed in measuring the MF viscosity. It is therefore often called the hydrodynamic volume V_h. The volume V_h may be by an order of magnitude higher than the magnetic volume V_f. An essential effect on the behavior of smaller particles used to obtain ferrofluids, is exerted by thermal fluctuations. In what follows, these particles will be called "cold" ones if this mechanism is disregarded.

Magnetic Moment Rotation in the Particle Body

The magnetic core of a particle is single-domain. Therefore, its magnetic moment $m = V_f M_s$ where M_s is the saturation magnetization of ferromagnetic material. Consider the characteristics which determine the magnetic moment dynamics in the body of the particle. In case of small crystalline-axis type anisotropy the energy of interaction of the magnetic moment with the external field and the particle body is determined by the relation

$$U = -\mu_0 \vec{m}\vec{H} - K_a V_f (\vec{m} \cdot \vec{n})^2/m^2 \tag{6.1}$$

Here K_a is the anisotropy constant, $\vec{n}$ the unit vector specifying the direction of the anisotropy axis. The qualitative nature of the magnetic moment rotations of a "cold" particle depends on the relationship between the terms in equality (6.1).

If $\mu_0 mH \ll K_a V_f$, the magnetic moment is rigidly bound with the axis of low magnetization, i.e., it is "frozen" into the body of the particle. Such a particle is referred to as a rigid dipole. In this case, rotation of the particle is the mechanism responsible for magnetic moment rotations. The magnetic moment is oriented along the low magnetization axis during attenuation of the Larmore magnetic moment processional motion

$$\tau_\gamma = (\beta\omega_\gamma)^{-1} \tag{6.2}$$

where $\omega_\gamma = \mu_0 \gamma H_a$ is the frequency of the ferromagnetic resonance equal in the order of magnitude to 10^9 Hz; $\gamma = 1.76 \times 10^{11}$ C/kg is the gyrometric ratio for an electron; β is the dimensionless attenuation parameter whose order of magnitude is 10^{-2}, $H_a = 2K_a/\mu_0 M_s$ is the anisotropic field intensity. The thermal rotational fluctuations of the moment decrease its association with the particle body. This mechanism is characterized by the dimensionless parameter

$$\sigma = K_a V_f / kT \tag{6.3}$$

equal to the anisotropy-to-thermal fluctuation energy ratio. Let us estimate this parameter. The anisotropy constant for ferromagnets ranges widely as $K_a \sim (10^6 - 10^2)$ J/M^3. At room temperatures $kT \sim 10^{-21}$ J and the particle radius a $\sim$ 4 nm, find $\sigma \sim 10^2 + 10^{-2}$. It is seen that σ may assume the values both greater and smaller than unity.

The development of equilibrium magnetic moment, averaged over the set of equal particles, is characterized by two times: the relaxation time, τ_0, of the magnetic moment at a certain direction of a bilateral low magnetization axis and the Neel relaxation time, τ_N, via the magnetic anisotropy barrier separating two equivalent magnetic moment directions:

$$\tau_0 = \tau_\gamma \sigma = m/(2\beta\gamma kT), \; \tau_N = \tau_\gamma \sigma^{-3/2} \exp\sigma \tag{6.4}$$

The range of τ_0 is only determined by the magnetic moment range, which is rather narrow for particles used in MF preparation. The characteristic value for $\tau_0 = 10^{-7}$ s. Therefore, in practically any hydrodynamic processes the averaged magnetic moment may be considered developed near a certain axis of magnetic anisotropy of a particle. The Neel time τ_N grows exponentially with increase and may range widely. The dynamic MF properties may therefore depend on the behavior of the Neel relaxation in particles used for its production.

Taking account of thermal fluctuations of the moment requires that the conditions of existence of rigid dipoles be ascertained. The fluctuation energy must be far lesser than the moment-particle body binding energy, i.e. $\sigma \gg 1$. However, even with this condition fulfilled, the Neel particle magnetization reversal occurs at any interval of about τ_N. The Neel particle may therefore be treated as a rigid dipole at an interval complying with the condition $t_* \ll \tau_N$.

If $\mu_0 m H \gg K_a V_f$, the orientation of the cold particle magnetic moment is always close to that of the magnetic field. In this case, which may be called a fixed moment approximation, the magnetic moment rotations are due to changes in the field orientation. As for a rigid dipole, the relaxation time may be estimated from relation (6.2) where the ferromagnetic resonance frequency is determined by the external field intensity. Thermal fluctuations stipulate stochastic rotations of the magnetic moment with respect to the field direction. The effect of this mechanism on the magnetic moment orientation is specified by the Langevin argument equal to the ratio of the, $\mu_0 m H$, energy of magnetic moment-field interaction to the energy of thermal fluctuations:

$$\xi = \mu_0 mH/kT = H/H_T \tag{6.5}$$

where the quantity $H_T = kT/\mu_0 m$ of the magnetic field intensity dimension is introduced from the condition $\xi = 1$. Estimate H_T. For magnetite particles with a characteristic volume $V_f = 5 \times 10^{-23}$ m^3 the magnetic moment $m = M_s V_f = 2.25 \times 10^{-17}$ A · m^2. For energy $kT = 4.15 \times 10^{-21}$ J, find $H_T = 1.46 \times 10^4$ A/m. $\xi \gg 1$ is an additional condition for the fixed moment approximation to be fulfilled, for in this case the fluctuational moment rotations are suppressed by the field.

In nonrigid cold dipoles, when the ratio $\mu_0 mH/kV_f$ may assume arbitrary values and the orientation of the vectors $\vec{H}$ and $\vec{n}$ is random, the magnetic moment is developed along the effective field

$$\vec{H}_e = -\partial U/\partial m = \vec{H} + \vec{H}_a \tag{6.6}$$

where $\vec{H}_a = \dfrac{2K_a}{\mu_0 M_s m}(\vec{m} \cdot \vec{n})\vec{n}$. The effect of thermal fluctuations on $\vec{m}$ orientation is specified by parameters σ and ξ.

Particle Rotation in a Viscous Fluid

The rotation of a particle body exerts a decisive effect on the magnetic moment relaxation in the hydrodynamic approximation. The concept of a rigid cold dipole is the simplest physical idealization that may be used for modeling the magnetic moment relaxation with regard to particle body rotation. Let us consider this model, as it allows the most simple formulation of the basic equation which may then be extended to allow for complicating factors.

The dynamics of rotational particle motion is described by the equation for its moment of momentum balance. The complete moment of momentum of the particle includes the moment of momentum of its rotational motion $I\vec{\omega}$ and the moment of momentum, $\gamma^{-1}\vec{m}$, of electrons participating in the formation of the magnetic moment. Here, $\vec{\omega}$ is the angular velocity of particle rotation; I the inertia moment equal to (2/5) $\rho V_h a^2$ for a spherical particle; a—the particle radius. The complete moment of the external forces, applied to the particle, is equal to the sum of moments of viscous friction forces $\alpha\vec{\omega}$ and magnetic forces $\mu_0 \vec{m} \times \vec{H}$ (α—is the rotational friction coefficient equal to $6V_h\eta_0$; η_0, the viscosity of fluid carrier). So, we have

$$\frac{d}{dt}(I\vec{\omega} + \gamma^{-1}\vec{m}) = -\alpha\vec{\omega} + \mu_0 \vec{m} \times \vec{H} \tag{6.7}$$

As follows from equation (6.7), rotary inertia exerts an essential effect on the particle dynamics if the time of variation of its rotational velocity is comparable with or less than the time of free rotation attenuation:

$$\tau_S = I/\alpha = \rho a^2/15\eta \tag{6.8}$$

For the characteristic values $\rho = 7500$ kg/m^3, $a = 10^{-8}$ m, $\eta = 10^{-3}$ kg/m · s we have the estimate $\tau_s = 5 \cdot 10^{-11}$ s. It is necessary to include the gyromagnetic effect if the time of magnetic moment rotation is comparable with or less than the attenuation time, τ_γ, of the magnetic moment precessional motion. These times are much shorter than the characteristic times of variation of macroscopic parameters, which allows the time variation of the complete moment of momentum in equation (6.7) to be disregarded (in the hydrodynamic approximation).

For a rigid bond of the magnetic moment and the particle body

$$\vec{\omega} = \vec{m} \times \frac{d\vec{m}}{dt} \cdot \frac{1}{m^2}$$

Taking account of this relation from (6.7) gives

$$(\alpha/m^2)\vec{m} \times (d\vec{m}/dt) = \mu_0 \vec{m} \times \vec{H} \tag{6.9}$$

In case of small deviations from the equilibrium when the angle between $\vec{m}$ and $\vec{H}$ is small, we may assume $\vec{m} \cdot \vec{H} \sim mH$. Including this relation, (6.9) takes on the form

$$d\vec{m}/dt = -\frac{1}{\tau}(\vec{m} - m\vec{e}) \tag{6.10}$$

where

$$\tau = \frac{\alpha}{\mu_0 mH} = \frac{6\eta}{\mu_0 M_s H} \frac{V_h}{V_f}, \quad \vec{e} = \frac{\vec{H}}{H}$$

In order to describe the magnetic moment dynamics, including the Brownian rotational diffusion, it is necessary to use kinetic equations. The magnetic moment relaxation equation, averaged by the set of equal rigid dipoles, has the form [14]

$$\frac{d}{dt} < \vec{m} > = -\frac{1}{\tau_\parallel}(< \vec{m} >_\parallel - m_0)\vec{e} - \frac{1}{\tau_\perp} < \vec{m} >_\perp \tag{6.11}$$

where

$$\tau_\parallel = \frac{\xi}{L}\frac{dL}{d\xi}\tau_B, \quad \tau_\perp = \frac{2L}{\xi - 1}\tau_B, \quad \tau_B = \frac{3V_h \eta_0}{kT} = \frac{1}{2}\tau\xi, \text{*}$$

$$L = \coth \xi - \frac{1}{\xi} \tag{6.12}$$

* See Nomenclature p. 211 for explanation of notation for trigonometric functions.

The subscripts $\|$ and $\perp$ correspondingly mean magnetic moment components parallel and normal to the field; $\tau_\|$, $\tau_\perp$, are the relaxation times of these component; τ_B is the Brownian time of rotational diffusion of particles. As seen from (6.11), the time variation of the averaged magnetic moment for Brownian dipoles is determined by relaxation, be it normal or longitudinal to the field of its components. Let us cite the asymptotic expressions for relaxation times in "week" ($\xi \ll 1$) and strong ($\xi \gg 1$) magnetic fields which follow from (6.12)

$$\left.\begin{aligned} &\tau_\| = \tau_\perp = \tau_B, && \xi << 1, \\ &\tau_\| = \tau_B/\xi = \frac{1}{2}\tau, \quad \tau_\perp = 2\tau_B/\xi = \tau, && \xi >> 1 \end{aligned}\right\} \tag{6.13}$$

In weak magnetic fields, the relaxation times are maximal and equal to τ_B. They decrease with the field intensity. In strong magnetic fields, relaxation time may be estimated through the relaxation time, τ, of the cold dipole. The characteristic value of the Brownian time is, as a rule, 10^{-5} to 10^{-6} s. Hydrodynamic times are much in excess of this value. By confining to low-frequency fields whose variation times may be compared with hydrodynamic ones, we can reduce equation (6.11). It includes a small parameter equal to the ratio of the maximum relaxation time $\tau_\|$, $\tau_\perp$ to the characteristic variation time $<\vec{m}>$ determined, in the case of fluid at rest, by the variation time of the magnetic field. In the zero approximation $<\vec{m}> = m_0\vec{e}$. In the first approximation, we shall confine to, obtain

$$<\vec{m}> = (m_0 - \tau_\| \frac{dm_0}{dH}\frac{dH}{dt})\vec{e} - \tau_\perp m_0 \frac{d\vec{e}}{dt} \tag{6.14}$$

The condition for equation (6.14) to be applied is

$$\omega_* \tau_\| \ll 1, \ \omega_* \tau_\perp \ll 1 \tag{6.15}$$

where ω_* is the characteristic frequency of field variation.

Expression (6.14) corresponds to the general form for the vector function dependent on one vector and linear in the time derivatives of the argument. As for a nonrigid dipole, suspended in the field at rest, the field vector is also the only external parameter which determines the averaged magnetic moment; it should be expected that the intraparticle relaxation mechanism may be included in equation (6.14) by specifying the scalar coefficients $\tau_\|$, $\tau_\perp$. Analysis of the rotational diffusion of nonrigid dipoles [43], based on the Fokker-Planck equation, provides an expression for the relaxation times at arbitrary σ and ξ ratio for two cases, $\tau_N \ll \tau_B$ and $\tau_N \gg \tau_B$. However, the expressions so derived are tedious and are omitted here. In what follows, an essential use is only made of the relaxation time dependence on the field intensity. This dependence is qualitatively correct in relations (6.12) and their simple asymptotics (6.13).

We shall conclude by citing the expression for the equilibrium distribution function used to detect the magnetic moment in the orientation space and the direction of low-magnetization axis of a nonrigid dipole [43]:

$$P_0 = Q_0^{-1} \exp\left[\xi \frac{\vec{m}\cdot\vec{e}}{m} + \sigma \frac{(\vec{m}\cdot\vec{n})^2}{m^2}\right]$$

Here Q_0 is the normalization integral. Using this distribution for the equilibrium value of the averaged moment part of equation (6.14), we obtain the Langevin relationship

$$m_0 = mL(\xi),\ L(\xi) = \coth\xi - \xi^{-1}* \tag{6.16}$$

6.2 MAGNETIZATION EQUATION FOR MAGNETIZATION IN LOW-FREQUENCY APPROXIMATION

Let us build the magnetization equation for magnetization. The magnetic moment of a material fluid element at rest differs from its equilibrium value only in a nonstationary magnetic field. The magnetization equation must therefore have its simplest form in the system of reference rigidly connected with the material element, relative to which it is at rest. In this system, the nonequilibrium part of magnetization is determined by partial time derivatives of the field components. The simplest assumption is that magnetization is only dependent on the first derivative components. The vector H derivative may be represented in the form

$$\frac{D\vec{H}}{Dt} = \frac{D(H\vec{e})}{Dt} = \frac{DH}{Dt}\vec{e} + H\frac{D\vec{e}}{Dt},\quad \vec{e} = \frac{\vec{H}}{H} \tag{6.17}$$

Here, the symbol D/Dt indicates that the derivative is calculated in the system of reference rigidly connected with the material element. The first term in the rhs of (6.17) is a vector directed along the field; the second one is normal to it. As the magnetization motion mechanisms may differ when only intensity of the field or only its direction is changing, these components of the vector $D\vec{H}/D\vec{t}$ will be considered as independent magnetization arguments, i.e., $\vec{M} = \vec{M}\left(\rho, T, \vec{H}, \frac{DH}{Dt}\vec{e}, H\frac{D\vec{e}}{Dt}\right)$. In a state of equilibrium, when $D\vec{H}/Dt = 0$, obtain $M = M_0(\rho, T, H)\vec{e}$. For a state close to equilibrium, by confining in the magnetization expansion to the linear terms over field vector derivative components, we get

* See Nomenclature p. 211 for explanation of notation for trigonometric functions.

$$\vec{M} = (M_0 - \kappa_{\parallel}\frac{DH}{Dt})\vec{e} - \kappa_{\perp} H \frac{D\vec{e}}{Dt} \tag{6.18}$$

or

$$\vec{M}' = \vec{M} - \vec{M}_0 = -\kappa_{\parallel}\frac{DH}{Dt}\vec{e} - \kappa_{\perp} H \frac{D\vec{e}}{Dt}$$

Here $\kappa_{\parallel} = \kappa_{\parallel}(\rho, T, H)$, $\kappa_{\perp} = \kappa_{\perp}(\rho, T, H)$ are the material coefficients specifying the dynamic part of magnetization $\vec{M}'$.

The system of reference, related to the material element, moves at linear velocity $\vec{v}$ and rotates at angular velocity $(1/2)\nabla \times \vec{v}$ with respect to the laboratory reference system. Transition to differentiation in the laboratory system is realized with the aid of the kinematic relations:

$$DH/Dt = dH/dt = \partial H/\partial t + (\vec{v}\cdot\nabla)H \ ; \ D\vec{e}/Dt = d\vec{e}/dt - (\tfrac{1}{2})(\nabla\times\vec{v})\times\vec{e} \tag{6.19}$$

Equation (6.18) is the simplest phenomenological hypothesis for magnetization dependence on dynamic parameters. It may be represented in an alternative from without vector $\vec{e}$:

$$\vec{M} = [M_0 - (\kappa_{\parallel} - \kappa_{\perp})\frac{dH}{dt}]\frac{\vec{H}}{H} + \kappa_{\perp}\frac{D\vec{H}}{Dt} \tag{6.20}$$

Now it is easy to proceed to the tensor statement

$$M_i = \chi H_i - \kappa_{ik}\left(\frac{D\vec{H}}{Dt}\right)_k \tag{6.21}$$

Here $\chi = M_0/H$ is the magnetic susceptibility. The tensor

$$\kappa_{ik} = \kappa_{\perp}\delta_{ik} - (\kappa_{\parallel} - \kappa_{\perp})H_i H_k/H^2 \tag{6.22}$$

characterizes the effect of dynamic processes on magnetization. By analogy with magnetic susceptibility, it may be referred to as a tensor of dynamic susceptibility, while the independent parameters $\kappa_{\parallel}$, $\kappa_{\perp}$, which determine its components, s coefficients of dynamic susceptibility.

Taking account of equation (6.14), it is easy to derive expressions for the coefficients of dynamic susceptibility in terms of microscopic characteristics of particles. Summing up Eq. (6.14) over all the particles in the unit volume gives an equation similar to (6.18). Comparing these equations, we obtain the following relations for a polydisperse fluid

$$\kappa_{\parallel} = \sum_i \chi_{ri}\tau_{\parallel i}, \ \kappa_{\perp} = \sum_i \chi_{si}\tau_{\perp i} \tag{6.23}$$

where $\chi_{si} = n_i m_{0i}/H$, $\chi_{ri} = n_i dm_{0i}/dH$ are the integral and differential magnetic susceptibilities, respectively; subscript i stands for the quantities of the i-th fraction; n_i is the number of particles of the i-th fraction per unit volume. For a monodisperse fluid from (6.23) including (6.12) we find

$$\kappa_{\parallel} = 3\kappa_0 \frac{\xi}{L}\left(\frac{dL}{d\xi}\right)^2; \quad \kappa_{\perp} = 6\kappa_0 \frac{2L^2}{\xi(\xi - L)} \tag{6.24}$$

Here $\kappa_0 = \varphi_h\eta_0/\mu_0 H_T$, $\varphi_h = n \cdot V_h$ is the hydrodynamic volume fraction of the particles. For limited cases of weak and strong fields we have:

$$\left.\begin{array}{ll} \kappa_{\parallel} = \kappa_{\perp} = \kappa_0 = \text{const} & \xi << 1 \\ \kappa_{\parallel} = 3\kappa_0/\xi^2, \quad \kappa_{\perp} = 6\kappa_0/\xi & \xi >> 1 \end{array}\right\} \tag{6.25}$$

As is seen, $\chi_{\parallel}$ and $\chi_{\perp}$ at $H \ll H_T$ may be considered to be constants independent of the field intensity. At $H \gg H_T$, they decrease as the field intensity increases, $\kappa_{\parallel}$ decreasing faster than $\kappa_{\perp}$.

In accordance with the structure of equation (6.18), the effects of dynamic interaction of MF and field may be divided into two groups where the relaxations of, correspondingly, longitudinal and normal magnetization components are predominant. The normal component is very significant in ferrohydrodynamics. It specifies the volume couples applied to the volume from the side of the field. It is precisely this mechanism that is responsible for a majority of effects to be considered later.

Subjecting equation (6.18) to vector multiplication by $\vec{H}$, obtain the expression for the density of magnetic couples

$$\vec{K} = \mu_0[\vec{M} \times \vec{H}] = \kappa_{\perp} \vec{H} \times D\vec{H}/Dt =$$

$$= 2\eta_r[2\vec{e} \times d\vec{e}/dt - \nabla \times \vec{v} + \vec{e}(\vec{e} \cdot \nabla \times \vec{v})] \tag{6.26}$$

Here, the coefficient of rotational viscosity is introduced

$$\eta_r = (1/4)\mu_0\kappa_{\perp}H^2 \tag{6.27}$$

which characterizes viscous friction forces that appear during relative rotation of the material element and the field. From (6.27) with allowance for (6.24), (6.25) we find

$$\eta_r = \eta_{rm} \frac{(\xi L^2)}{(\xi - L)}$$

at $\xi \ll 1$

$$\eta_r = (1/6)\eta_{rm}\xi^2 \tag{6.28}$$

at $\xi \gg 1$

$$\eta_r = \eta_{rm}(1 - \xi^{-1})$$

Here, $\eta_{rm} = (3/2)\varphi_h\eta_0$ is the rotational saturation viscosity for rigid dipoles achieved in strong magnetic fields ($\xi \to \infty$). Rotational saturation viscosity of nonrigid dipoles depends on the parameter σ. Various expressions for this relation are obtained in [14, 43]. In the limited cases of small and large σ, they result in

$$\eta_{rs} = \eta_{rm} \begin{cases} \frac{2}{45}\sigma^2 & \text{at } \sigma \ll 1 \\ 1 - \frac{1}{2\sigma} & \text{at } \sigma \gg 1 \end{cases} \tag{6.29}$$

It is seen that, with increasing σ, the coefficient of rotational viscosity in a state of saturation grows to achieve its maximum η_{rm} for rigid dipoles. Otherwise, when $\sigma \ll 1$, the coefficient of rotational viscosity and the normal magnetization component are vanishing. In this case the condition of parallel vectors M and H is fulfilled and used in the model of fluid with equilibrium magnetization.

The idea of magnetic fluids as systems of noninteracting particles provides estimates for the coefficient of rotational viscosity for fluids of low concentration. At moderate and high concentrations, an essential effect on the volume density of magnetic couples may be exerted by particle interaction. The dependence of the coefficient of rotational saturation viscosity on the hydrodynamic fraction of particles φ_h at moderate φ_h may fairly be approximated by the expressions

$$\eta_{rs} = (3/2)\varphi_h\eta\,, \quad \eta = \eta_0 \exp\left(\frac{2{,}5\varphi_h + 2{,}7\varphi_h^2}{1 - 0{,}609\varphi_h}\right) \tag{6.30}$$

where η is the fluid viscosity coefficient without the field. In Fig. 6.1, the line plots the dimensionless parameter $S = \eta_{rs}/\eta$ vs the volumetric concentration of solid phase $\varphi_s = nV_s$ (obtained from density measurements) in accordance with the first relation (6.30) at $\rho = \varphi_h/\varphi_s = 2.5$. The circles show the experimental results obtained by measuring the magnetoviscous effect in kerosene- and magnetite-based MFs. The value of p was estimated by processing viscometric measurements, with the field off. In Fig. 6.2, the lines furnish the logarithm of the MF-to-kerosene viscosity ratio η_0 vs. volumetric concentration of the solid phase at different values of p plotted according to the Wand formula (second relation (6.30)) for viscosity of concentrated suspensions.

6.3 FORMULATION OF A CLOSED SET OF EQUATIONS

Let us formulate the set of hydrodynamic equations for a magnetizable conductive uncharged and electrically nonpolarizable fluid taking account of the dy-

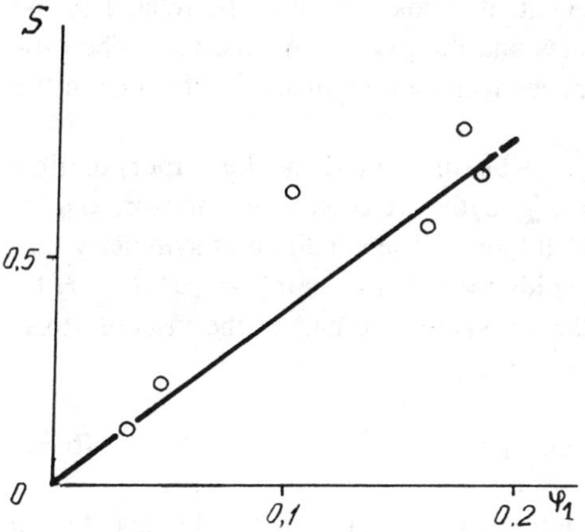

FIG. 6.1

namic magnetization component. We shall assume that at any instant of time magnetization does not greatly differ from its equilibrium value, which is the case in low-frequency approximation (6.15). The set is based on general balance equations of mass, momentum, heat and moment-of-momentum transfer which for a homogeneous one-phase medium are, correspondingly, of the form

$$\begin{aligned} \dot{\rho} &= \rho \nabla \cdot \vec{v}; \\ \rho(\dot{\vec{v}})_i &= \nabla_k \sigma_{ik} + \rho g_i , \\ \rho T \dot{s} &= -\nabla \vec{q} + \psi , \\ \epsilon_{ikl} \sigma_{kl} &= 0 \end{aligned} \tag{6.31}$$

Here, σ_{ik} is the complete tensor of medium (field) stress; s the entropy of a unit fluid mass; $\vec{q}$ the heat flux; ψ the dissipative function. Disregarding the internal moment of momentum of the fluid, the balance equation of the moment of momentum is reduced to the condition of symmetry of the complete stress tensor.

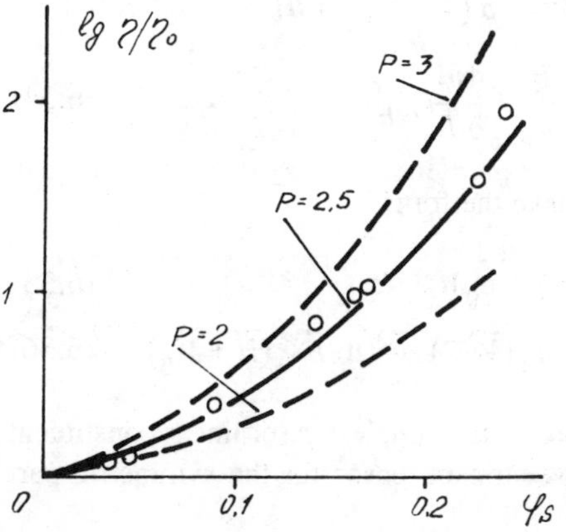

FIG. 6.2

In magnetic fluids, the internal moment of momentum may be related to the rotary inertia of ferromagnetic particles and the gyromagnetic effect. The estimates cited in 6.1 show that these mechanisms are of small significance in the low-frequency approximation.

The complete stress tensor may be presented as the superposition $\sigma_{ik} = -p\delta_{ik} + \sigma'_{ik} + \sigma^e_{ik}$, where σ'_{ik} is the viscous stress tensor; $\sigma^e_{ik} = (-1/2)\mu_0 H^2\delta_{ik} + H_i B_k$ is the Maxwell tensor. The condition of symmetry σ_{ik}, with allowance for this expansion, yields the relation $\epsilon_{ikl}\sigma'_{ik} = \mu_0(\vec{M} \times \vec{H})_l$, whence follows the expression for the antisymmetric part of the viscous stress tensor

$$\sigma'^a_{ik} = (1/2)\mu_0\epsilon_{ikl}(\vec{M} \times \vec{H})_l \tag{6.32}$$

Thus, the volume source in momentum balance equation (6.31), due to the fluid-field interaction, is $\vec{F}_i = \nabla_k(\Delta p\delta_{ik} + \sigma'^a_{ik} + \sigma^e_{ik})$. Here, Δp is the contribution to the pressure specified by the medium magnetization. Taking account of the field equations for the source, obtain the expression

$$\vec{F} = -\nabla(\Delta p) + (1/2)\mu_0\nabla \times (\vec{M} \times \vec{H}) + \mu_0(\vec{M}\nabla)\vec{H} + j \times \vec{B} \tag{6.33}$$

Fluid magnetization at any instant of time may be represented as a sum of the equilibrium part $\vec{M}_0 = M_0(\rho, H, T)\vec{H}/H$ corresponding to instantaneous values of the thermodynamic parameters and the dynamic (dissipative) part $\vec{M}'$ due to nonequilibrium processes, i.e., $\vec{M} = \vec{M}_0 + \vec{M}'$. In accordance with such division, the volume magnetic source $\vec{F} = \vec{F}_0 + \vec{F}'$, magnetization contribution to pressure $\Delta p = \Delta p_0 + \Delta p'$ and entropy $\Delta S = \Delta S_0 + \Delta S'$, may also be represented as superposition. In the lowfrequency approximation, when the dissipative component of magnetization is small, it is natural to ignore its effect on the thermodynamic parameters, i.e., to set $\Delta p' = \Delta S' = 0$. In this case, the contribution of magnetization to pressure and entropy may be calculated by the relations used in quasi-stationary ferrohydrodynamics

$$\Delta p = \Delta p_0 = \mu_0 \int_0^H (M_0 - \rho(\frac{\partial M}{\partial \rho})_{T,H})dH$$
$$\Delta s = \Delta s_0 = \rho^{-1}\mu_0 \int_0^H (\frac{\partial M}{\partial T})_{\rho,H}dH \tag{6.34}$$

The magnetic source, F, components have the form

$$\vec{F}_0 = -\nabla(\Delta p) + \mu_0 M_0 \nabla H \tag{6.35}$$

$$\vec{F}' = (1/2)\mu_0\nabla \times (\vec{M}' \times \vec{H}) + \mu_0(\vec{M}'\cdot\nabla)\vec{H} + \mu_0\vec{j} \times (\vec{H} + \vec{M}_0) \tag{6.36}$$

To close the basic balance equations in the simplest case, linear constituent equations may be used, such as the Newtonian tensor for the symmetric part of the viscous stress tensor $\sigma_{ik}'^{S} = (\eta_v - 2\eta/3)\nabla\cdot\vec{v}\,\delta_{ik} + \eta(\nabla_i v_k + \nabla_k v_i)$, the Fourier equation for a heat flux $\vec{q} = -\lambda\nabla T$, Ohm's law for current density $j = \sigma\cdot(\vec{E} + \vec{v}\times\vec{B})$, and equation (6.18) for the dynamic part of magnetization.

As the dynamic magnetization component is called forth by dissipative processes, the heat transfer equation should include heat release due to this process. In the first approximation the dissipative function ψ is quadratic in the independent parameters determining nonequilibrium processes. In case of a dissipative magnetization component, they are $\dot{H}$ and $H(\dot{e} - 0.5(\nabla\times\vec{v})\times\vec{e})$. Thus, for dissipative heat release due to magnetization relaxation we obtain

$$\psi_M = \mu_0\kappa_\parallel \dot{H}^2 + \mu_0\kappa_\perp H^2(\dot{\vec{e}} - 0.5(\nabla\times\vec{v})\times\vec{e})^2 =$$
$$= -\mu_0\vec{M}'(\vec{H} - 0{,}5(\nabla\times\vec{v})\times\vec{H})$$

Taking account of this relation for the complete dissipative function due to viscosity, electrical conductivity and magnetization relaxation, we have the expression

$$\psi = 0.5\eta(\nabla_i v_k + v_i\nabla_k)^2 + (\eta_v - 2\eta/3)(\nabla\cdot\vec{v})^2 +$$
$$+ \sigma^{-1}j^2 - \mu_0\vec{M}'(\vec{H} - 0.5(\nabla\times\vec{v})\times\vec{H}) \tag{6.37}$$

The above relations make it possible to formulate a closed set of ferrohydrodynamics equations. Let us cite it for an incompressible nonconducting fluid in case of constant dynamic viscosity and thermal conductivity coefficients with neglect of the magnetocaloric effect

$$\rho\dot{\vec{v}} = -\nabla p + \eta\nabla^2\vec{v} + 0.5\mu_0\nabla\times(\vec{M}\times\vec{H}) + \mu_0(\vec{M}\cdot\nabla)\vec{H} + \rho\vec{g} \tag{6.38}$$

$$\rho c_p\dot{T} = \lambda\nabla^2 T + 0.5\eta(\nabla_i v_k + \nabla_k v_i)^2 - \mu_0\vec{M}'\cdot(\vec{H} - 0.5(\nabla\times\vec{v})\times\vec{H})) \tag{6.39}$$

$$\nabla\cdot\vec{v} = 0, \quad \nabla\cdot\vec{B} = 0, \quad \nabla\times\vec{H} = 0 \tag{6.40}$$

where

$$\vec{M}' = -\Delta\kappa\dot{H}^{\vec{H}}/H - \kappa(\vec{H} - 0.5(\nabla\times\vec{v})\times\vec{H}),\ \vec{M} = \vec{M}_0 + \vec{M}'$$
$$\Delta\kappa = \kappa_\parallel - \kappa_\perp\quad \kappa\equiv\kappa_\perp,\ \vec{B} = \vec{H} + \vec{M}_0 + \vec{M}' \tag{6.41}$$

The conditions of continuity of the velocity $[\vec{v}] = 0$ and normal component of the magnetic induction vector $[\vec{H} + \vec{M}_0 + \vec{M}']\vec{n} = 0$, discontinuity of the tangential component of the field intensity vector $[\vec{H}]\times\vec{n} = \vec{j}_s$ and total stress vector $[\vec{\sigma}] = \vec{\sigma}_s$ at the interface of two media with different properties, may be used as boundary conditions to solve this set of equations. Here, $\vec{j}_s$, $\vec{\sigma}_s$ are the density of surface currents and nonmagnetic forces (for example, surface tension forces); n is the normal to the interface.

Let us dwell in greater detail upon the last of the boundary conditions. The total stress vector (whose components $\sigma_i = \sigma_{ik}n_k$) is specified by the relation

$$\vec{\sigma} = -(1/2)\mu_0 H^2\vec{n} + B_n\vec{H} + (1/2)\mu_0\vec{n} \times (\vec{M}' \times \vec{H}) + \vec{\sigma}'^s \quad (6.42)$$

where $\sigma_i'^s = \sigma_{ik}'^s n_k$. Carry out scalar and vector multiplication of the condition for the total stress vector discontinuity by $\vec{n}$, taking account of (6.42) and the boundary conditions for the field vectors, to obtain conditions for normal and tangential stress discontinuities, respectively

$$[(1/2)\mu_0((\vec{M}_0 + \vec{M}')\cdot\vec{n})^2 + \vec{\sigma}^s\vec{n}] + (1/2)\mu_0(\vec{j}_s \times (\vec{H}_1 + \vec{H}_2))\cdot\vec{n} = \vec{\sigma}_s\cdot\vec{n} \quad (6.43)$$

$$[(1/2)\mu_0(\vec{M}\times\vec{H}) - (1/2)\mu_0(\vec{n}\cdot\vec{M}\times\vec{H})\vec{n} + \vec{\sigma}^s\times\vec{n}] + \vec{j}_s(\vec{B}\cdot\vec{n}) = \vec{\sigma}_s\times\vec{n} \quad (6.44)$$

Here H_1 and H_2 are the values of the vector H from different sides of the interface.

Let us cite the basic balance equations for MFs in an integral form. Momentum balance equation (6.38) for a nonconducting fluid may be presented in the form

$$\rho(\dot{\vec{v}})_i = \nabla_k(-p\delta_{ik} + \sigma_{ik}') + M_k\nabla_k H_i \quad (6.45)$$

Here $\sigma_{ik}' = \sigma_{ik}'^s + \sigma_{ik}'^a$. After integrating this equation over the fluid volume and transforming the divergence volume integral to the surface one, we shall arrive at the equation

$$\frac{d}{dt}\int_V \rho\, v\, dV = \int_S (\vec{\sigma}' - p\vec{n})\, dV + \mu_0 \int_V (\vec{M}\nabla)\vec{H}\, d\cdot S \quad (6.46)$$

Here,

$$\vec{\sigma}' = \vec{\sigma}'^s + \vec{\sigma}'^a, \quad \vec{\sigma}'^a = (1/2)\mu_0\vec{n}\times(\vec{M}'\times\vec{H}) \quad (6.47)$$

Subjecting equation (6.45) to vector multiplication by $\vec{r}$, we obtain the equation

$$\rho(\vec{r}\times\dot{\vec{v}})_i = \nabla_k(\epsilon_{ijl}r_j(-p\delta_{lk} + \sigma_{lk}')) + \mu_0(\vec{M}\times\vec{H})_i + (\vec{r}\times\mu_0(\vec{M}\cdot\nabla)\vec{H})_i \quad (6.48)$$

After integration, hence follows the integral balance equation for the moment of momentum

$$\frac{d}{dt}\int_V \rho(\vec{r}\times\vec{v})dV = \int_S \vec{r}\times(\vec{\sigma}' - p\vec{n})dS + \mu_0\int_V (\vec{M}\times\vec{H})dV + $$
$$+\mu_0\int \vec{r}\times(\vec{M}\cdot\nabla)\vec{H}dV \tag{6.49}$$

Noninductive Approximation

The equations of motion and magnetic field in the set (6.38) through (6.41) are interrelated, which greatly complicates its solution. However, a wide range of problems may be simplified by omitting dynamic variables from the Maxwell equations. For simplicity, let us find out the condition of applicability of such an approximation for weak magnetic field when, by virtue of (6.25) and (6.23) for a monodisperse fluid we have

$$\vec{M}' = -\chi_0\tau_B D\vec{H}/Dt \tag{6.50}$$

Here χ_0 and τ_0 are constants. Taking account of this relation, the dynamic source in the equation of motion and Maxwell equations may be written in the form

$$\vec{F}' = \mu_0\chi_0\tau_B\,(0.5\nabla\times(\vec{H}\times\frac{D\vec{H}}{Dt}) - (\frac{D\vec{H}}{Dt}\cdot\nabla)\vec{H}) \tag{6.51}$$

$$\nabla\times\vec{H} = 0, \quad \nabla\cdot(\vec{H} + \chi_0(\vec{H} - \tau_B D\vec{H}/Dt)) \tag{6.52}$$

In the low-frequency approximation, the characteristic times of field variation are high as compared to the time of magnetization relaxation and, hence, the dimensionless operator $\tau_B DH/Dt$ small. Besides, consider the case when $\chi_0 \ll 1$, which is often fulfilled in practice. Thus, expressions (6.51) and (6.52) contain two small parameters. If we confine in (6.51) to a quadratic small value approximation, we should ignore in it the magnetic field induced by the magnetic fluid. A similar approximation should be used to calculate the surface source due to the dynamic part of magnetization in boundary condition (6.44). The field in the noninductive approximation satisfies the equations

$$\nabla\times\vec{H} = 0, \nabla\cdot\vec{H} = 0 \tag{6.53}$$

As magnetic susceptibility, χ_0, for MFs in weak fields has a maximum value, the condition $\chi_0 \ll 1$ is the condition for applicability of the noninductive approximation in arbitrary fields.

CHAPTER

SEVEN

MAIN EFFECTS OF VOLUME MAGNETIC COUPLES IN A UNIFORM FIELD

7.1 EQUATION OF MOTION FOR MAGNETIC FLUID IN A UNIFORM FIELD

For a magnetic fluid moving in a uniform external field, the volume magnetic force $\mu_0(\vec{M}\nabla)\vec{H}$ in a noninductive approximation should be set zero for any flow geometry. However, in the equation of motion, a part of the magnetic source due to the action of couples is other than zero. From (6.26), taking account of the continuity equation, we find

$$(1/2)\nabla \times \vec{K} = \eta_r \nabla^2 \vec{v} + \eta_r \nabla \times \vec{e}(\vec{e} \cdot \nabla \cdot \vec{v}) \tag{7.1}$$

which allows the equation of motion to be written in the form

$$\rho d\vec{v}/dt = -\nabla p + (\eta + \eta_r)\nabla^2 \vec{v} + \eta_r \nabla \times \vec{e}(\vec{e} \cdot \nabla \times \vec{v}) + \rho \vec{g} \tag{7.2}$$

Equation (7.2) is applicable both to the stationary and arbitrary time-dependent uniform magnetic field. In this case, the components of the space-constant vector $\vec{e}$ and η_r are time-dependent.

In order to formulate particular problems, we should use expressions for the operators in Eq. (7.2) in curvilinear orthogonal coordinate systems. As is seen, this equation differs from the Navier-Stokes equation in the term containing the vector $\vec{e}$. In the Navier-Stokes equations, the operators in different coordinate systems are widely reported in literature. Here, we shall discuss how the unit vector $\vec{e}$, specifying equation (7.2), is prescribed.

The field orientation relative to the Cartesian coordinate system may appropriately be prescribed in terms of the direction cosines e_x, e_y, e_z related by

$$e_x^2 + e_y^2 + e_z^2 = 1 \tag{7.3}$$

The angles α and β, for example, may be used as independent parameters (Fig. 7.1). The direction cosines are related with the independent parameters as

$$e_x = \sin\alpha \cdot \cos\beta,\ e_y = \sin\alpha \cdot \sin\beta,\ e_z = \cos\alpha \tag{7.4}$$

The problem symmetry often allows the axis z to be chosen so that $\beta = 0$. Then relations (7.4) are reduced.

The field rotating in some plane may be prescribed in the form

$$\vec{e} = \vec{e}_0 \cos\omega t + \vec{e}_1 \sin\omega t \tag{7.5}$$

where $\vec{e}_0$, $\vec{e}_1$ are the field orientations at some initial instant of time and at phase $\omega t = \pi/2$, respectively. The components of the vectors $\vec{e}_0$ and $\vec{e}_1$ are expressed in terms of the independent parameters α and β which determine the direction of the unit normal $\vec{n}$ to the plane of rotation (Fig. 7.2) prescribing the direction of the angular rotational velocity vector. For an initial vector $\vec{e}_0$ the orientation of $\vec{e}$ is appropriate when it is in the plane between the axis z and vector $\vec{n}$. In this case, for the direction cosines of the vectors $\vec{e}_0$ and $e_1 = n \times e_0$ we obtain the relations

$$\left.\begin{aligned} &e_{0x} = \pm\cos\alpha\cos\beta\ ,\ e_{0y} = \pm\cos\alpha\sin\beta\ ;\ e_{0z} = \mp\sin\alpha\ ; \\ &e_{1x} = \mp\sin\beta\ ,\ e_{1y} = \cos\beta\ ,\ e_{1z} = 0 \end{aligned}\right\} \tag{7.6}$$

Here, the upper sign is chosen for $0 < \alpha < \pi/2$, the lower for $\pi/2 < \alpha < \pi$.

Let us consider the prescription of the unit vector $\vec{e}$ in a cylindrical coordi-

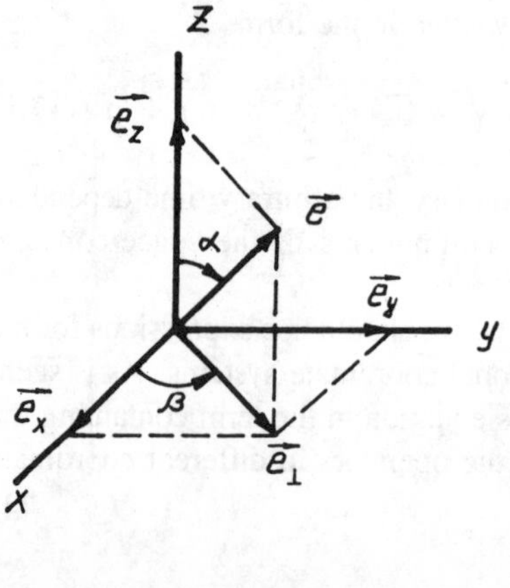

FIG. 7.1

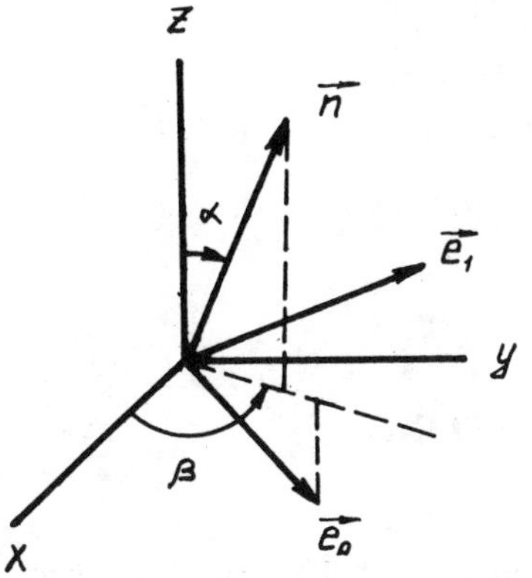

FIG. 7.2

nate system. As two independent parameters, choose the angle α, which constitutes unit vector $\vec{e}$ with the axis z, and angle β between the projection of $\vec{e}$ onto the plane normal to the axis z and the origin of the azimuthal angle φ. Then, the unit vector $\vec{e}$ projections onto the orthogonal datum point of the cylindrical coordinate system at an arbitrary point have the form

$$e_r = \sin\alpha\cos(\beta - \varphi),\ e_\varphi = \sin\alpha\sin(\beta - \varphi);\ e_z = \cos\alpha \tag{7.7}$$

Consider a spherical coordinate system. Let the direction cosines of the vector $\vec{e}$ at the initial datum point $\vec{i}_{r0}, \vec{i}_{\varphi 0}, \vec{i}_{\theta 0}$, i.e., at $\theta = \varphi = 0$ are equal to $e_{r0}, e_{\theta 0}, e_{\varphi 0}$. Similarly to (7.4), they may be expressed in terms of two independent angles. The components of the vector $\vec{e}$ at an arbitrary point of the sphere will be obtained by making a transition to the coordinate system turned relative to the axis $\vec{i}_{r0}$ by the angle φ and then relative to the axis $\vec{i}_{\varphi 0}$ by the angle θ. As a result, obtain

$$\left.\begin{aligned} e_r &= e_{r0}\cos\theta + (e_{\theta 0}\cos\varphi + e_{\varphi 0}\sin\varphi)\sin\theta\,; \\ e_\theta &= -e_{r0}\sin\theta + (e_{\theta 0}\cos\varphi + e_{\varphi 0}\sin\varphi)\cos\theta\,; \\ e_\varphi &= -e_{\varphi 0}\sin\varphi + e_{\varphi 0}\cos\varphi \end{aligned}\right\} \tag{7.8}$$

Let us conclude with the equation of motion (7.2) in the projections onto the axis which will be most often used in the Cartesian coordinate system:

$$\left.\begin{aligned} &\rho dv_x/dt = -\partial p/\partial x + (\eta + \eta_r)\nabla^2 v_x + 2\eta_r\,(e_z\partial\Omega_e/\partial y - \\ &- e_y\partial\Omega_e/\partial z) + \rho g_x\,; \\ &\rho dv_y/dt = -\partial p/\partial y + (\eta + \eta_r)\nabla^2 v_y + 2\eta_r\,(e_x\partial\Omega_e/\partial z - e_z\partial\Omega_e/\partial x) + \\ &+ \rho g_y\,; \\ &\rho dv_z/dt = -\partial p/\partial z + (\eta + \eta_r)\nabla^2 v_z + 2\eta_r\,(e_y\partial\Omega_e/\partial x - \\ &- e_x\partial\Omega_e/\partial z) + \rho g_z \end{aligned}\right\} \tag{7.9}$$

where

$$d/dt = \partial/\partial t + v_x \partial/\partial x + v_y \partial/\partial y + v_z \partial/\partial z \ ;$$

$$\nabla^2 = \partial^2/\partial x^2 + \partial^2/\partial y^2 + \partial^2/\partial z^2 \ ;$$

$$2\Omega_e = (\partial v_z/\partial y - \partial v_y/\partial z)e_x + (\partial v_x/\partial z - \partial v_z/\partial x)e_y +$$
$$+ (\partial v_y/\partial x - \partial v_x/\partial y)\, e_z$$

7.2 VISCOUS STRESSES IN FLUID RESTING OR ROTATING AS SOLID

Equation (7.2) assumes steady rotation of a fluid as solid. This case is the simplest one for a distinct dynamic interaction.

Consider a vessel of arbitrary form which rotates at a constant angular velocity $\vec{\Omega}$ in an external uniform field rotating at an angular velocity $\vec{\omega}$. The vector $\vec{\omega}$ may change in time obeying an arbitrary law. In the noninductive approximation, the field inside the fluid may be considered uniform. For ellipsoidal vessels the fluid is also uniform with regard to the fields induced by MF magnetization.

In a fluid at rest or rotating as solid, there are no symmetric viscous stresses and the viscous stress tensor is antisymmetric: $\sigma'_{ik} = \sigma'^{a}_{ik} = \dfrac{1}{2}\,\epsilon_{ikl}K_l$. The force of viscous stresses acting on unit area of vessel, $\vec{\sigma}\,' = \dfrac{1}{2}\vec{n} \times \vec{K}$, is tangential to the surface.

Figures 7.3 and 7.4 illustrate the mechanism of antisymmetric viscous stresses, correspondingly, for a fluid at rest in a rotating field and for a fluid rotating in a stationary field. Here and further on, for simplicity we shall regard particles in our microscopic consideration as rigid dipoles. As is seen from Fig. 7.3, the particles rotating lagging the field set the layer adjacent to the interface into a shear state responsible for viscous stresses acting upon the wall.

Figure 7.4 shows three successive positions of a particle at the wall of the rotating vessel. The circular motion of the particle is translational as its orientation is fixed by the stationary field. Because of this, orientation of the particle with respect to the adjacent interface section is changed and a shear flow in the boundary layer is induced. The investigator rotating with the vessel interpretes

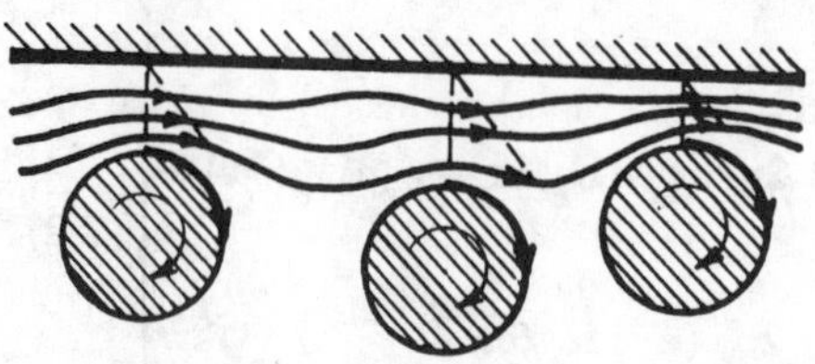

FIG. 7.3

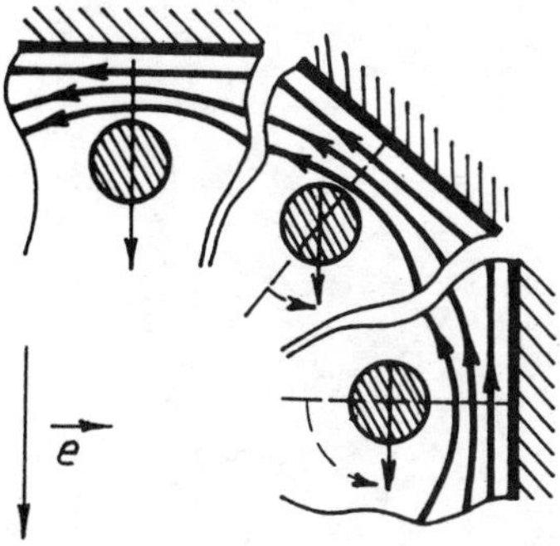

FIG. 7.4

antisymmetric stresses in accordance with Fig. 7.3. In the laboratory system, the stationary field seems to him to be rotating at a velocity equal in modulus and opposite in direction to the velocity of vessel rotation.

As $\vec{K}$ is a constant, the total force affecting the entire surface is equal to zero. Tangential viscous stresses may be registered by measuring the rotational moments. It follows from (6.44) that the dynamic component of the total rotational moment in the noninductive approximation

$$\vec{\mathcal{K}} = \mu_0 \int_V (\vec{M} \times \vec{H})\, dV = \vec{K} \cdot V$$

where V is the fluid volume. Thus, the dynamic rotational moment, in this case, is determined by the product of the magnetic couple density by the volume. Let us study the dependence of the couple density vector on the orientation of the vectors $\vec{e}$, $\vec{\Omega}$ and $\vec{\omega}$. By assuming $(1/2)\nabla \times \vec{v} = \vec{\Omega}$, $\vec{e} \times d\vec{e}/dt = \omega$ in relation (6.26), obtain

$$\vec{K} = 4\eta_r[\vec{\omega} - \vec{\Omega}(1 - e_z) + \Omega e_z \vec{e}_\perp] \tag{7.10}$$

Here, $\vec{e}_z$, $\vec{e}_\perp$ are, correspondingly, the components of the unit vector $\vec{e}$ parallel and perpendicular to the angular vessel velocity $\vec{\Omega}$.

In a variable magnetic field, the unit vector $\vec{e}$ components and the rotational viscosity coefficient η_r and, hence, $\vec{K}$, are time-dependent. Consider the case when a constant-intensity field rotates at a velocity constant in modulus and angular in direction. As follows from (7.5), the vector $\vec{e}$ components may now be presented in the form

$$e_z = e_{0z}\cos\omega t + e_{1z}\sin\omega t; \ \vec{e}_\perp = \vec{e}_{0\perp}\cos\omega t + \vec{e}_{1\perp}\sin\omega t$$

where $\vec{e}_0$, $\vec{e}_1$ are time-independent unit vectors. Let us consider the value of $\vec{K}$ mean for one revolution of the field. If at the initial instant of time the field direction, prescribed by the unit vector $\vec{e}$, is chosen so that it lies in the plane of vectors $\vec{\Omega}$ and $\vec{\omega}$, then the component e_{1z} equals to zero and the mean pulsating values in (7.10) are:

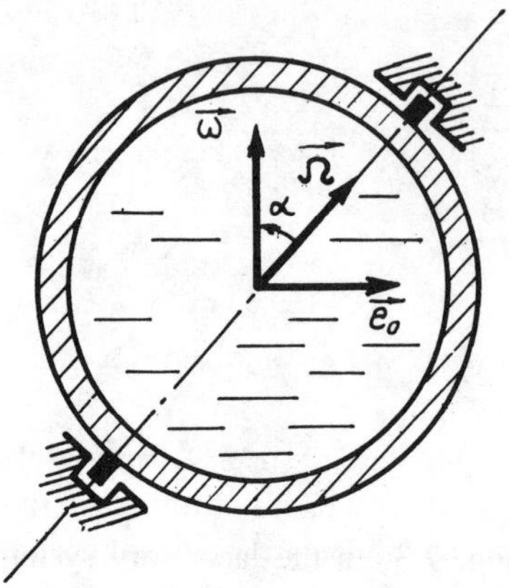

FIG. 7.5

$$
\begin{aligned}
(e_z^2)_m &= (1/2)e_{0z}^2 = (1/2)\sin^2\alpha; \\
(e_z\vec{e}_\perp)_m &= (1/2)e_{0z}\vec{e}_{0\perp} = (1/4)\sin 2\alpha \cdot \vec{i}
\end{aligned}
\tag{7.11}
$$

where α is the angle between the vectors $\vec{\Omega}$ and $\vec{\omega}$, $\vec{i}$ is a unit vector in the $\vec{e}_{0\perp}$ direction. Taking account of these relations, find

$$
\vec{K}_m = 4\eta_r[\vec{\omega} - \vec{\Omega}(1 - 0.5e_{0z}^2) + (1/2)\Omega e_{0z}\vec{e}_{0\perp}]
\tag{7.12}
$$

As seen from (7.10) and (7.12), the direction of the rotational moment does not, in the general case, coincide with the direction of the vessel rotation axis. For the instantaneous and mean values of the axial and transverse components of the couple density we have

$$
\left.
\begin{aligned}
K_z &= 4\eta_r[\omega_z - \Omega(1 - e_{0z}^2]; \; K_{zm} = 4\eta_r[\omega_z - \Omega(1 - 0.5e_{0z}^2)]; \\
\vec{K}_\perp &= 4\eta_r(\vec{\omega}_\perp + \vec{\Omega}e_z\vec{e}_\perp); \; \vec{K}_{\perp m} = 4\eta_r(\vec{\omega} + 0.5\Omega e_{0z}\vec{e}_{0\perp})
\end{aligned}
\right\}
\tag{7.13}
$$

Assuming in (7.10) and in all subsequent relations, successively, that $\omega = 0$ and $\Omega = 0$, we obtain density for magnetic couples for a fluid rotating in a stationary field and a fluid at rest in a rotating field, respectively.

Let us consider a specific example. First, we shall estimate the rotation velocity of the MF vessel, whose axis of rotation is fixed and friction in the supports is eliminated by the rotating field. Here, the vector of the angular velocity of field rotation may not coincide with the direction of the vessel rotation axis. Because of zero friction $K_{zm} = 0$ and from (7.13) it follows

$$
\Omega = \frac{2\omega_z}{2-e_{0z}^2} = \frac{2\omega\cos\alpha}{1+\cos^2\alpha}
\tag{7.14}
$$

For the second equality, it is considered that, according to (7.6), $e_{0z} = \sin\alpha$. From the second equality (7.13) and taking account of (7.14) we find the transverse component of the density of magnetic couples

$$\vec{K}_{\perp m} = 4\eta_r\omega \frac{\sin\alpha}{1 + \cos^2\alpha} \frac{e_{0\perp}}{e_{0\perp}} \tag{7.15}$$

As seen, this component is only equal to zero when the directions of vectors $\vec{\Omega}$ and $\vec{\omega}$ coincide.

Let us extend relations (7.14) and (7.15) to the law of linear friction in the supports. The condition of equilibrium of the axial component of rotational moments is of the form $K_{zm} = r\,\Omega$, where r is the coefficient of rotational friction in the supports. Hence, with account of (7.13), we find

$$\Omega = \frac{2\omega\cos\alpha}{1 + 2\hat{r} + \cos^2\alpha}, \quad K_{\perp m} = 4\eta_r\omega \frac{(1 + 2\hat{r})\sin\alpha}{1 + 2\hat{r} + \cos^2\alpha} \tag{7.16}$$

Here $\hat{r} = r/(4\eta_r V)$ is the dimensionless parameter.

Measurements of the rotational moments allow the rotational viscosity coefficient to be estimated immediately from equation (7.10) and relations following from (7.10).

7.3 EFFECTS OF MAGNETIC COUPLES AT SIMPLE SHEAR

The effects of dynamic interaction which may take place in shear flows can be illustrated in the simplest form for a simple shear, a plane-parallel flow with a linear velocity profile $v_x = 2\Omega_y$. This profile satisfies equation (7.2) at random orientation of the magnetic field with respect to the coordinate axis. The velocity curl $\vec{\Omega} = (1/2\nabla) \times \vec{v}$ is directed along the axis z. Viscous stress force (6.47), acting per unit area of the interface normal to the axis y, is a sum of symmetric and antisymmetric viscous stresses to be

$$\vec{\sigma}_y = -2\eta\Omega\vec{i}_x + 0.5[\vec{i}_y \times \vec{K}] \equiv \sigma_y \tag{7.17}$$

Here $\vec{i}_x$, $\vec{i}_y$ are the unit vectors of the Cartesian coordinate system. The magnetic couple density is related by (7.10), where $\vec{\Omega} = \Omega\vec{i}_z$. As a result, we find

$$\vec{\sigma}_y = -2[\eta + \eta_r(1 - e_z^2)]\Omega\vec{i}_x + 2\eta_r(\vec{i}_y \times \vec{\omega} + \Omega e_z \vec{i}_y \times \vec{e}_\perp)] \tag{7.18}$$

Let us consider some immediate consequences that follow from this equation.

Magnetoviscous Effect

For the component of viscous tangential stresses, acting in the direction of fluid motion in a stationary field, we have

$$\sigma_{xy} = -2\eta_e\Omega \tag{7.19}$$

where $\eta_e = \eta + \eta_r(1 - e_z^2)$ is the effective MF viscosity in the magnetic field. With the magnetic field off, $\eta_r = 0$ and (7.19) becomes the known relation for a normal liquid, $\sigma_{xy} = -2\eta\Omega$. With the field on, the rotational viscosity mechanism starts operating, which can be observed as growing effective viscosity. The increment $\eta_e - \eta = \eta_r(1 - e_z^2)$, referred to as magnetoviscous effect, is determined by the rotational viscosity coefficient and field orientation with respect to the macroscopic velocity curl. Let us introduce the dimensionless parameter characterizing a relative change of the longitudinal component of viscous stresses, with the field on (relative magnetoviscous effect):

$$S_{\|} = (\eta_e/\eta) - 1 = S(1 - e_z^2) = S\sin^2\alpha \tag{7.20}$$

Here $S = \eta_r/\eta$; α is the angle between the macroscopic curl (axis z) and the vector $\vec{e}$. As seen, S is linearly dependent on the parameter S at any field orientations. This is because the magnetic field does not distort the velocity profile of a simple shear flow. The maximum value of the relative magnetoviscous effect is attained when the field is in the plane normal to the shear plane ($e_z = 0$) and $S_{\|} = S$.

The mechanism of the magnetoviscous effect is furnished in Fig. 7.6. Particle orientation at three successive instants of time is shown at the lower boundary. The dashed arrow shows orientation of the particle when no magnetic field is applied, when the particle is freely rotating in a shear flow. The solid arrow shows the particle in a transverse field, when its orientation is fixed. The velocity profile of a microscopic shear flow due to decelerated particle rotation is shown in the boundary layer. It is seen that the microscopic and main (at the upper wall) shears are similarly directed, which increases viscous friction at the wall with the field on.

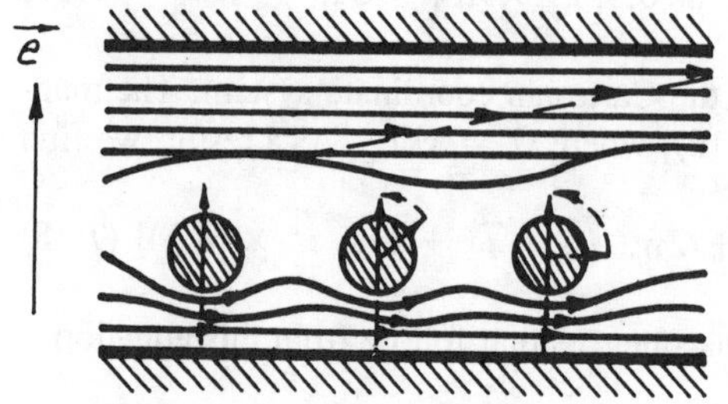

FIG. 7.6

Transverse Stresses

As follows from (7.18), viscous stresses have a component normal to the fluid direction. In a stationary field it is equal to

$$\sigma_{zy} = -2\eta_r e_x e_z \Omega \tag{7.21}$$

The value of σ_{zy} is maximal when the field lies in the shear plane ($e_y = 0$) and its direction, together with the fluid direction, constitute the angle $\pi/4$ ($e_x = e_z = \sqrt{2}/2$). Here $(\sigma_{zy})_{max} = -\eta_r \Omega$.

As a dimensionless parameter characterizing transverse stresses, we can choose the ratio of z- and x-components of the viscous stress vector (equal to the tangent between the vector of total stresses and fluid direction):

$$S_\perp = \frac{\sigma_{zy}}{\sigma_{xy}} = \frac{s e_x e_z}{1+S(1-e_z^2)} \tag{7.22}$$

Find orientation of the field at which $S_\perp$ attains its maximum value. From the condition $\partial S_\perp / \partial e_z = 0$ obtain stationary values of the direction field cosine and axis z

$$e_{zm} = \sqrt{\frac{(1-e_y^2)(1+S)}{2+S(1+e_y^2)}} \tag{7.23}$$

at which $S_\perp$ attains its maximum value if the domain of field orientations forms a circular conic surface with its axis normal to the shear plane

$$S_{\perp m} = \frac{S(1-e_y^2)}{2\sqrt{(1+S)(1+Se_y^2)}} \tag{7.24}$$

If the field is in the shear plane, $S_\perp$ attains its absolute maximum at all possible changes of the field orientation. Setting in (7.23) and (7.24) $e_y = 0$, we arrive at

$$e_{z*} = \sqrt{\frac{1+S}{2+S}}, \quad S_{\perp *} = \frac{S}{2\sqrt{1+S}} \tag{7.25}$$

In the approximation for diluted fluids, when $S \ll 1$, we have from (7.25)

$$e_{z*} = \frac{\sqrt{2}}{2}\left(1+\frac{S}{4}\right), S_{\perp *} = \frac{1}{2}S \tag{7.26}$$

As follows from (7.20) and (7.26), maximum value of the relative parameter of transverse stresses in diluted fluids is equal to the relative magnetoviscous effect at the same field orientation, i.e., $S_{\perp *} = S_\parallel$.

The mechanism of the transverse stress effect is illustrated in Fig. 7.7 where a set of parallel lines shows the current lines of the shear flow on the plane z, x. Shown in the right-hand top corner is the diagram of angular velocities in which Ω is the macroscopic curl; $\vec{\omega}$ the rotational velocity of the particle whose axis is fixed by the field in the direction of the unit vector $\vec{e}$ in the plane z, x. The vector $\vec{\Omega}$ falls into two components which are longitudinal and normal to the vector $\vec{e}$. The field completely retards the normal component of the angular particle velocity and does not affect the longitudinal one. The vector $\vec{\omega}$ is therefore sought as a longitudinal component of the vector $\vec{\Omega}$. If the vector $\vec{e}$ is inclined to the macroscopic curl, $\vec{\omega}$ has components that are parallel and normal to $\vec{\Omega}$: $\omega_\parallel = \Omega e_z^2$, $\omega_\perp = \Omega e_z e_x$; here, $\omega_\parallel < \Omega$. The difference between $\Omega - \omega_\parallel$ and $\omega_\perp$ is specified by microscopic shear flows in the boundary layer which cause antisymmetric viscous stresses. The vector diagram of viscous stresses, affecting the upper plane, is presented in the left-hand bottom corner in Fig. 7.7, where $\vec{\sigma}^s$ is the vector of viscous stresses, with the field off; $\vec{\sigma}^a$ is the vector of antisymmetric viscous stresses caused by the field. The component $\sigma_\parallel^a = \sigma_{xy}^a$ is due to retardation of the particle angular rotational velocity component longitudinal to the macroscopic curl and proportional to the difference $\Omega - \omega_\parallel$. The component $\sigma_\perp^a = \sigma_{xy}^a$ is specified by the generation with a shear flow of the particle angular velocity component $\omega_\perp$ which disappears when the field is off.

Compensation of Viscous Stresses in a Shear Flow

Consider a simple shear with an applied rotating magnetic field. For simplicity assume that the angular rotational velocity of the field and the macroscopic velocity curl are directed along the same line, i.e., field orientation changes in the plane normal to the shear plane. In this case, as it follows from (7.18), the viscous stress tensor component for the force acting in the fluid direction is

$$\sigma_{xy} = -2(\eta + \eta_r)\Omega + 2\eta_r\omega \tag{7.27}$$

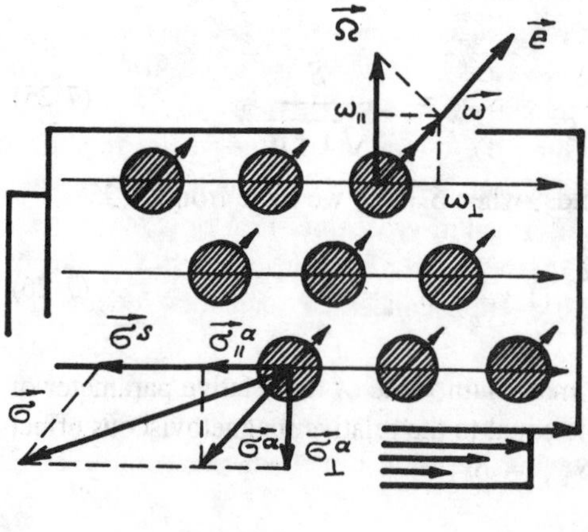

FIG. 7.7

It is seen from (7.27) that a low-frequency ($\omega \rightarrow 0$) rotating field increases viscous stresses, i.e., a magnetoviscous effect takes place. Higher rotational frequency either increases still further or decreases it depending on the signs of ω and Ω. If the signs of Ω and ω are different, the absolute value of viscous stresses grows with $|\omega|$. If the signs are the same, $|\sigma_{xy}|$ decreases and, at $\omega = \Omega$, the magnetoviscous effect is compensated: the field does not affect viscous friction. The latter is only determined by the dynamic viscosity mechanism. A further increase of ω decreases $|\sigma_{xy}|$. When the condition

$$\Omega = \frac{\eta_r \omega}{\eta + \eta_r} = \frac{S\omega}{1+S} = S_1\omega \tag{7.28}$$

is satisfied, viscous stresses disappear and then, with growing $|\omega|$, change their sign. In this case, viscous friction forces promote, rather than inhibit, relative motion of the boundaries until the macroscopic velocity curl satisfies condition (7.28).

The microscopic meaning of the viscous stress compensation effect in a shear flow is as follows. The shear rate of the secondary flow in the boundary layer due to particle-field interaction may have a sign opposite to the macroscopic shear rate. In this case, the resultant shear rate at the boundary decreases with viscous stress force. In case of complete compensation, the macroscopic and microscopic shear rates are equal in their magnitude and opposite in their sign, the resultant shear rate being zero.

It should be noted that the vanishing of the viscous stress tensor component for the shear resistance force is not identical to disappearance of viscous dissipation. As follows from (6.37), for the case under consideration, the energy dissipated per unit volume is

$$\psi = \eta(dv/dy)^2 + 4\eta_r(\omega - \Omega)^2 = 4\omega^2\eta_r\eta/(\eta + \eta_r) \tag{7.29}$$

The rotating field is not a factor necessary for the observation of compensation of viscous stresses. Generally speaking, this effect takes place if at a fixed direction of the macroscopic shear there is a mechanism which may reverse the sign of the microscopic shear rate. In case of a stationary uniform field, fluid rotation, for instance, is such a mechanism. As follows from Fig. 7.4, the microscopic shear reverses its sign when the direction of fluid vessel rotation is changed. Therefore, should the direction of vessel rotation be appropriately chosen, viscous stresses in some prescribed macroscopic shear flow in the vessel will be attenuated and completely compensated. The particulars of this mechanism will be discussed in the next section concerned with the Couette flow between rotating cylinders.

Fluid motion in a nonuniform magnetic field provides other opportunities for observing the effect. As a particle is travelling in a nonuniform field, its magnetic moment is rotating following the field direction, irrespective of the macroscopic shear state of the fluid. The direction of this rotation and, consequently, the direction of microscopic shear are determined by the particle direction. Thus, at a fixed macroscopic shear velocity, viscous stresses may be compensated by changing the direction of particle velocity.

Therefore, the effect of viscous stress compensation may show itself diversely. As an example, let us discuss the formation of a zero-rate and even ascending fluid flow on an inclined plane in a rotating uniform field. Consider a plane-parallel flow in a layer of thickness l with a free boundary. The problem geometry is shown in Fig. 7.8. Projection of the steady-state equation of motion onto the axis x may be written in the form

$$d\sigma'_{xy}/dy = -\rho g \sin\alpha \tag{7.30}$$

Taking account of the free-boundary conditions $\sigma'_{xy}|_{y=l} = 0$, find from (7.27) and (7.30)

$$\rho g \sin\alpha (l - y) = -(\eta + \eta_r) dv/dy + 2\eta_r \omega \tag{7.31}$$

Hence, with regard to the rigid boundary condition $v|_{y=0} = 0$, we have the velocity distribution and fluid flow rate

$$v = -\frac{2\eta_r}{\eta + \eta_r}\omega y + \frac{\rho g \sin\alpha}{\eta + \eta_r} y\left(l - \frac{1}{2}y\right);$$
$$Q = \frac{Fl}{\eta + \eta_r}\left(\frac{1}{3}\rho g \sin\alpha - \eta_r \omega\right) \tag{7.32}$$

where F is the film cross-sectional area. The qualitative form of velocity profile (7.32) is depicted in Fig. 7.8. If the direction of field rotation is such that the particles, rotating behind the field, roll down the inclined plane, the field accelerates the flow-down of the fluid; otherwise, fluid motion is retarded and, at

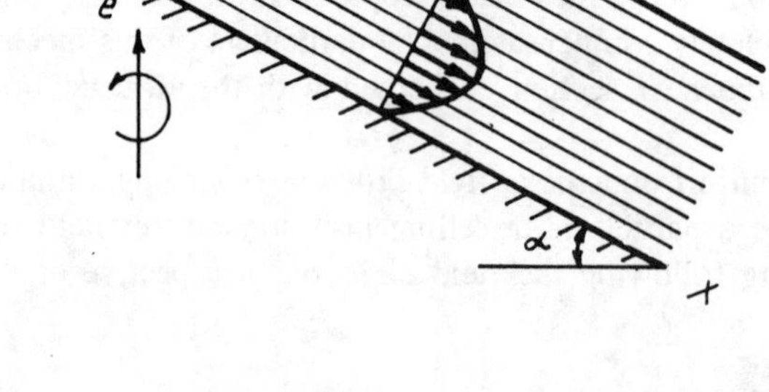

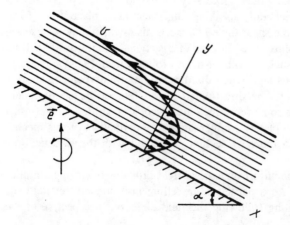

FIG. 7.8

$$\rho g l \sin\alpha = 3\eta_r\omega \tag{7.33}$$

zero-rate flow sets in. The gravity flow is balanced by the flow due to the particles rolling up the inclined plane. In the resultant stream, the flow near the rigid boundary is descending, while it is ascending at the free boundary. It is worth noting that dynamic equilibrium is only determined by rotational viscosity independent of shear viscosity. Estimate the thickness of the plane-parallel zero-rate film on a vertical surface. Assuming $\rho = 10^3$ kg/m^3; $\eta_r = (3/2)\eta_0 = 10^{-3}$ kg/(m · s); $\omega = 314$ Hz, find $l = 0.1$ mm. From (7.32) obtain the condition under which velocity on the surface reverses the sign

$$\rho g l \sin\alpha = 4\eta_r\omega \tag{7.34}$$

In this case, the effect of complete compensation (vanishing) of viscous stresses in a shear flow can be observed near the free film surface.

Thus, dynamic interaction between the MF and the field (interaction via three-dimensional couples for a uniform field) is reduced mainly to three effects which are magnetoviscous effect, compensation of longitudinal and generation of transverse stresses. The following sections of the Chapter concern these effects for the most important one-dimensional flows.

7.4 THE COUETTE FLOW IN AN ANNULAR GAP

Let us consider a flow between two infinite coaxial nonmagnetic cylinders. Let the internal cylinder of radius R_1 be rotating at angular velocity Ω_1, while the external one of radius R_2 at velocity Ω_2. The uniform field is rotating at angular velocity ω. Here, the direction $\vec{\omega}$ may not coincide with the cylinder axis.

As the velocity curl Ω for a normal viscous fluid flow between rotating cylinders is constant, it is clear that the Couette velocity profile also satisfies Eq. (7.2) at a random orientation of the magnetic field. So, just as for a viscous fluid, the velocity profile of the flow under consideration is of the form

$$v = \Omega r - DR_2^2/r \tag{7.35}$$

where

$$\Omega = \frac{\Omega_2 - \Omega_1 a^2}{1 - a^2}\,;\quad D = \frac{\Omega_2\Omega_1}{1 - a^2}\,;\quad a = \frac{R_1}{R_2} \tag{7.36}$$

v is the azimuthal velocity component in a cylindrical coordinate system; Ω is the macroscopic velocity curl; D is the maximum shear rate achieved near the internal cylinder.

For the tangential stress vector we have from (6.49) the expression

$$\vec{\sigma}'_r = 2\eta D \frac{R_1^2}{r^2} \vec{i}_\varphi + \frac{1}{2}[\vec{i}_r \times \vec{K}] \tag{7.37}$$

Here, $\vec{i}_r$ and $\vec{i}_\varphi$ are the radial and azimuthal unit vectors of the cylindrical coordinate system, respectively. The instantaneous and mean values (for a rotating field of the magnetic couple density in (7.37) are specified by relations (7.10) and (7.12) where the macroscopic velocity curl $\vec{\Omega} = \vec{\Omega} i_z$. Due to constant Ω, the magnetic couple density is also a constant vector through the flow. The viscous stress component due to dynamic viscosity, as can be seen from (7.37), varies over the gap width.

The rotational moments of the viscous forces acting upon the cylinders in case of the Couette flow, are measured experimentally. Through vector multiplication of (7.37) by $\vec{r}$, integration over the cylinder perimeter for the axial component of the moments per unit length of the internal and external cylinders, and by making necessary transformations, we shall, correspondingly, obtain the following equations

$$\mathcal{K}_{z1} = \frac{4\pi\eta R_1^2}{1-a^2}\{\Omega_2 - \Omega_1 - S[(\omega_z - \Omega_2 e_\perp^2)(1-a^2) - (\Omega_2 - \Omega_1)a^2 e_\perp^2]\};$$

$$\mathcal{K}_{z2} = -\frac{4\pi\eta R_1^2}{1-a^2}\{\Omega_2 - \Omega_1 - \frac{S}{a^2}[\omega_z - \Omega_2 e_\perp^2)(1-a^2) - (\Omega_2 - \Omega_1)a^2 e_\perp^2]\} \tag{7.38}$$

Here, $e_\perp$ is the unit vector $\vec{e}$ component normal to the axis z.

It can easily be seen that the moments affecting the internal and external cylinders are different because of the moment of momentum of the magnetic field source in the fluid volume. Following (6.48), the rotational moment $\mathcal{K}_z$ affecting the field source, when in interaction with the fluid per unit length of the cylinder, is an algebraic sum of the moments applied to the cylinders taken with an opposite sign

$$\mathcal{K}_z = -(\mathcal{K}_{z1} + \mathcal{K}_{z2}) = 4\pi\eta_r R_2^2[(\omega_z - \Omega_2 e_\perp^2)(1 - a^2) - (\Omega_2 - \Omega_1)a^2 e_\perp^2] \tag{7.39}$$

Let us analyze relations (7.38) and (7.39).

Magnetoviscous Effect

The magnetoviscous effect in a stationary magnetic field may be estimated by measuring the rotational moment on a cylinder at rest. In this case, (7.38) yields the expressions

$$\mathcal{K}_{z1.2} = \frac{4\pi R_1^2 \Omega_{2.1}}{1-a^2}\eta(1 + Se_\perp^2) \tag{7.40}$$

Compare (7.40) with the similar equality for a viscous fluid ($\eta_r = 0$) to find the expressions for effective viscosity and relative magnetoviscous effect which are identical to (7.19) and (7.20):

$$\eta_e = \eta(1 + Se_\perp^2),\ S_\parallel = S(1 - e_z^2) \tag{7.41}$$

If both cylinders and the field are in rotation, fluid motion includes, as a constituent element, quasi-solid fluid rotation relative to the field. As rotational viscosity is able to give rise to viscous stresses in fluids rotating as solids, while dynamic viscosity is deprived of this ability, the field effect, in this case, cannot be reduced to the magnetoviscous effect.

Viscous Stress Compensation in a Shear Flow

Consider the effects due to compensation of viscous stresses when the field rotates in the plane normal to the cylinder axes. Here, the rotational moments have only the axial component. Their expressions follow from (7.38) and (7.39) where it is necessary to set $\omega_z = \omega$, $e_\perp = 1$. If the angular rotational velocity, ω, of the field is equal to the fluid velocity curl, Ω, in modulus and direction, the rotational viscosity mechanism disappears. The rotational moments are only determined by dynamic viscosity. Taking account of expression (7.36) for Ω, the condition of magnetoviscous effect compensation may be presented in the form

$$W = \frac{\omega - \Omega_2}{\Omega_2 - \Omega_1} = \frac{a^2}{1 - a^2} \tag{7.42}$$

Here, a dimensionless parameter W is introduced for the rotational velocity of the cylinders and the field, which is always positive since $a < 1$. As follows from (7.39), the fluid and the field source do not interact in this case.

The identity $\mathcal{K}_{z1} = 0$ gives the condition under which viscous stresses on the internal cylinder disappear.

$$W = \frac{\omega - \Omega_2}{\Omega_2 - \Omega_1} = \frac{1 + Sa^2}{S(1 - a^2)} \tag{7.43}$$

Here, the external cylinder is affected by the moment equal in modulus and opposite in sign to the moment applied to the magnetic field source:

$$\mathcal{K}_{z2} = -\mathcal{K} = 4\pi\eta R_2^2(\Omega_2 - \Omega_1) = 4\pi\eta_r R_2^2 \frac{1 - a^2}{1 + Sa^2}(\omega - \Omega_2) \tag{7.44}$$

Relation (7.43) may be used for estimating the parameter S by the known parameter W:

$$S = [W - a^2(1 + W)]^{-1} \tag{7.45}$$

Relation (7.44) may serve for independent estimations of dynamic and rotational viscosities.

Let us consider viscous manifestations of the effect when viscous stresses on the internal cylinder disappear. If a stationary cylindrical vessel with MF and with an unfastened internal cylinder is placed in a rotating field, the cylinder will start rotating. For this case $W = -\omega/\Omega_1$. As W is always greater than zero, the sign " $-$ " indicates that the cylinder and the field are rotating in opposite directions.

If the vessel is set in motion in a stationary field, $W = -\Omega_2/(\Omega_2 - \Omega_1)$. As $W > 0$, then $\Omega_2 < \Omega_1$, i.e., the unfastened cylinder rotates at a velocity higher than the rotational velocity of the vessel.

The equality $\mathcal{K}_{z1} = 0$ yields the condition under which viscous stresses on the external cylinder disappear

$$W = \frac{\Omega - \Omega_2}{\Omega_2 - \Omega_1} = \frac{(1+S)a^2}{S(1-a^2)} \tag{7.46}$$

The rotational moment on the internal cylinder is determined by the relation identical to (7.44).

Transverse Stresses

For the flow between rotating cylinders, transverse stresses are directed along the cylinder axes. They give rise to a rotational moment which strives to alter the cylinder axis orientation in space. The expressions for the normal component of the rotational moment per unit length of the internal and external cylinders are of the form

$$\vec{\mathcal{K}}_{\perp 1,2} = -\frac{2\pi\eta_r R_{1,2}^2}{1-a^2}[\vec{\omega}_\perp (1-a^2) + (\Omega_2 - \Omega_1 a^2)e_z \vec{e}_\perp] \tag{7.47}$$

Consider the case when transverse stresses are measured on a cylinder at rest in a stationary field. From (7.47) we have

$$\mathcal{K}_{\perp 1.2} = \mp \frac{2\pi\eta_r R_1^2}{1-a^2}\Omega_{2.1} e_z \vec{e}_\perp \tag{7.48}$$

These relations imply that the disturbing rotational moment is maximum when the field and the cylinder axis form an angle $\pi/4(e_z = e_\perp = \sqrt{2}/2)$. As a dimensionless parameter for the transverse stress effect in the Couette flow, we may choose the tangent of an angle between the vector of the complete rotational moment and the cylinder axis. It is

$$S_{\perp} = -\frac{Se_z e_{\perp}}{2[1 + S(1 - e_z^2)]} \tag{7.49}$$

From the condition $\partial S_{\perp}/\partial e_z = 0$ find the value of the direction field cosine to the axis z

$$e_{z*} = \sqrt{\frac{(1+S)}{(2+S)}} \tag{7.50}$$

at which $S_{\perp}$ attains its maximum

$$S_{\perp *} = S/4\sqrt{1 + S} \tag{7.51}$$

Relations (7.50) and (7.51) may be compared with relations (7.25) for maximum transverse stresses at a simple shear. It is seen that orientation of the field relative to the axis z, at which maximum is attained, is the same in both cases. The maximum value of the relative parameter of transverse stresses in case of the Couette flow is twice as small as that for a simple shear.

7.5 POISEUILLE FLOW IN AN ELLIPTICAL CHANNEL

Let us consider a steady-state flow in a long elliptical channel caused by a pressure drop in uniform field of random orientation relative to the channel axis. Choose the coordinate system as follows: the axis z is along the channel; the axes x and y are along the axes of the flow ellipse. The velocity profile is specified by equation (7.2) and the boundary conditions

$$v = 0 \text{ at } x^2/a^2 + y^2/b^2 = 1 \tag{7.52}$$

where a, b are the lengths of the flow ellipse semi-axes. A steady-state solution of equations (7.52) is to be tried in the form of a linear flow $v_z = v(x,y)$. For such a profile, the projections of equation (7.2) onto the axis z and the plane normal to the axis z have the form

$$\begin{aligned} \partial p/\partial z &= (\eta + \eta_r e_z^2)\nabla^2 v + \eta_r (\vec{e}\cdot\nabla)^2 v; \\ \nabla_{\perp} p &= \eta_r e_z [\vec{e}_{\perp}\nabla^2 v - \nabla(\vec{e}\cdot\nabla) v] \end{aligned} \tag{7.53}$$

Here, $\nabla_{\perp} p = \vec{i}_x \partial p/\partial x + \vec{i}_y \partial p/\partial y$ is the projection of the pressure gradient onto the cross-sectional plane of the channel (transverse pressure gradient);

$$\vec{e}_{\perp} = e_x \vec{i}_x + e_y \vec{i}_y = \vec{e}_x + \vec{e}_y; \; (\vec{e} \cdot \nabla) = e_x \partial/\partial x + e_y \partial/\partial y$$

Solution of problem (7.52), (7.53) to be tried as a Newtonian flow profile of the prescribed geometry

$$v = v_m(1 - x^2/a^2 - y^2/b^2) \tag{7.54}$$

Substitution of (7.54) into (7.53) and the necessary calculations give

$$\frac{\partial p}{\partial z} = -2\, v_m\, \eta_e\, \frac{(a^2 + b^2)}{a^2 b^2}; \quad \nabla_\perp p = -2\, v_m\, \eta_r\, \frac{a^2 \vec{e}_x + b^2 \vec{e}_y}{a^2 + b^2} \tag{7.55}$$

where

$$\eta_e = \eta + \eta_r \frac{[\,(1 - e_x^2)a^2 + (1 - e_y^2)b^2]}{(a^2 + b^2)} \tag{7.56}$$

Magnetoviscous Effect

A relation similar to the first equality in (7.55) is satisfied for a viscous fluid (i.e., with the field off). With the field on, the effective viscosity η_e, specified by relation (7.56), acts as a dynamic viscosity coefficient. By setting $a = b$ in (7.56), we obtain the expression for effective viscosity for a circular channel flow

$$\eta_e = \eta + (1/2)\eta_r(1 + e_z^2) = \eta + (1/2)\eta_r(1 + \cos^2\alpha) \tag{7.57}$$

where α is the angle between the field direction and the channel axis. For a plane-parallel channel whose plane is parallel to the plane $[x, z]$, from (7.56) at $b/a \to 0$ follows the relation

$$\eta_e = \eta + \eta_r(1 - e_x^2) \tag{7.58}$$

Let us represent (7.56) in a dimensionless form. For the relative magnetoviscous effect obtain the expression

$$S_\parallel = \frac{\eta_e}{\eta} - 1 = S\, \frac{(1 - e_z^2)a^2 + (1 - e_y^2)b^2}{a^2 + b^2} \tag{7.59}$$

Expressions for the relative magnetoviscous effect in the field longitudinal and transverse relative to the channel axis that are usually realized in experiments, are presented below:

$\vec{e} =$	$\vec{i}_x$	$\vec{i}_y$	$\vec{i}_z$
General case	$\dfrac{b^2 S}{a^2+b^2}$	$\dfrac{a^2 S}{a^2+b^2}$	S

Circular channel	$\frac{1}{2}S$	$\frac{1}{2}S$	S
Plane-parallel channel	0	S	S

As is seen, the magnetoviscous effect in a circular channel in the longitudinal field is twice as large as that in a transverse channel.

Transverse Stresses

Transverse stresses, distributed unevenly over the channel cross section, occur in a fluid flowing in the field inclined to the channel axis. They cause a volume force normal to the channel axis. For an axial flow this volume force is calculated by the rhs of the second equation in (7.53). For profile (7.54), this force is constant in modulus and direction throughout the fluid at any field orientations and is balanced by the transverse pressure gradient, as follows from the second relation (7.55). Thus, in channel flows due to the pressure gradient, transverse stresses show themselves as pressure drop in the channel cross-sectional plane.

For a circular channel we find from (7.55)

$$\nabla_\perp p = \frac{2\eta_r e_z \vec{e}_\perp}{2\eta + \eta_r (1 + e_z^2)} \cdot \frac{dp}{dz} \tag{7.60}$$

It is seen that, in this case, the transverse pressure gradient is directed along the field vector component normal to the channel axis. It is in this direction that the transverse pressure drop on the channel walls will be maximum. As a dimensionless parameter for transverse stresses we may choose the tangent of the angle between the total pressure gradient and the channel axis. It is

$$S_\perp = \frac{2Se_z e_\perp}{2 + S(1 + e_z^2)} \tag{7.61}$$

In longitudinal and transverse fields $S_\perp = 0$. Hence, there is some intermediate field orientation prescribed by the direction cosine e_{z*} at which $S_\perp$ is maximum. From the condition $\partial S_\perp / \partial e_z = 0$ and from relation (7.61) find

$$e_{z*} = \sqrt{\frac{2 + S}{4 + 3S}}; \quad S_{\perp *} = \frac{S}{2\sqrt{2(S + 1)(S + 2)}} \tag{7.62}$$

In the approximation for diluted MFs when $S \ll 1$, we find from (7.62)

$$e_{z*} = (1 - S/4)\sqrt{2}/2; \; S_\perp = S/4 \tag{7.63}$$

It is seen that, at small values of S, the relative transverse pressure gradient is maximum in the fields at an angle of 45° to the channel axis. In this case, as follows from (7.59), it is equal to one-third of the relative magnetoviscous effect $S_{\perp *} = S_{\|*}/3$.

Let us consider the characteristics of the transverse pressure gradient for a plane-parallel channel flow. From (7.55) at $b/a = 0$ we have

$$\nabla_{\perp} p = \frac{\eta_r e_z \vec{e}_x}{\eta + \eta_r (1 - e_x^2)} \frac{dp}{dz}; \quad S_{\perp} = \frac{S e_z e_x}{1 + S(1 - e_x^2)} \tag{7.64}$$

The expression for S is identical to relation (7.22) obtained for a simple shear accurate to the choice of the coordinate axes. Therefore, for the parameter $S_{\perp}$ in case of the Poiseuille plane-parallel channel flow, equalities (7.23) through (7.26) hold as obtained for a simple shear.

Note that in spite of the existence of the transverse pressure gradient, there is no resultant transverse force of dynamic interaction of the channel and the field source. This conclusion follows from equation (6.46). The force due to the transverse pressure gradient is balanced by the transverse component of the vector of the antisymmetric stresses applied to the channel walls.

7.6 PROPAGATION OF VISCOUS PERTURBATIONS

Let us consider the characteristics of viscous one-dimensional perturbations. Such perturbations may, for example, be caused by an infinite flat plate performing harmonic oscillations in its plane at angular frequency. Direct the y axis of the Cartesian coordinate system normal to the plate, and the x axis along the oscillations. The problem geometry is depicted in Fig. 7.9. Solution of equation

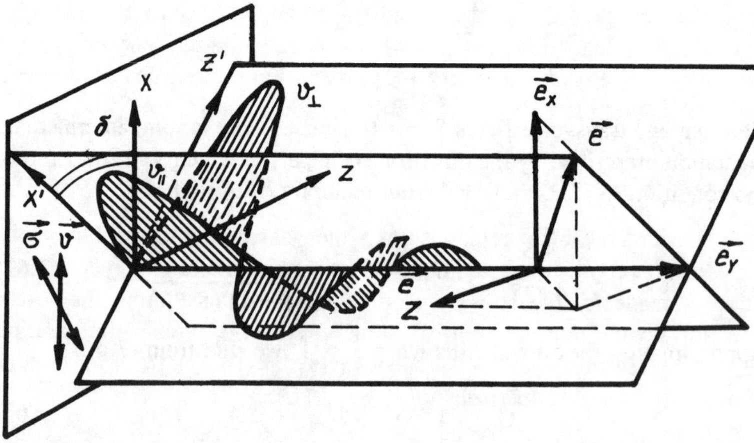

FIG. 7.9

(7.2) is to be tried in the form of transverse perturbations whose characteristics change only along the axis y: $\vec{v} = \{v_x(y,t),\ 0,\ v_z(y,t)\}$. The projections of equation (7.2) onto the plane parallel to the plate are of the form

$$\left.\begin{aligned} \partial v_x/\partial t &= [\eta + \eta_r\,(1 - e_z^2)\,]\,\partial^2 v_x/\partial y^2 + \eta_r e_x e_z\,\partial^2 v_z/\partial y^2\,; \\ \partial v_z/\partial t &= [\eta + \eta_r\,(1 - e_x^2)]\partial^2 v_z/\partial y^2 + \eta_r\,e_x e_z \partial^2 v_x/\partial y^2 \end{aligned}\right\} \qquad (7.65)$$

The set of equations (7.65) is essentially reduced in the hatched coordinate system in which the field vector has only two components. This may be achieved by turning the initial coordinate system about the axis y. Suppose that the coordinate system is turned so that the field is in the plane $\{x',\ y\}$. In this system $e'_z = 0.1 - e_{x'}^2 = e_y^2$, and the set of equations (7.65) takes on the form

$$\partial v_{\parallel}/\partial t = \eta_{\parallel}\,\partial^2 v_{\parallel}/\partial y^2\ ;\ \partial v_{\perp}/\partial t = \eta_{\perp}\,\partial^2 v_{\perp}/\partial y^2 \qquad (7.66)$$

where

$$\eta_{\parallel} = \eta + \eta_r\ ,\ \eta_{\perp} = \eta + \eta_r\,e_y^2\,; \qquad (7.67)$$

$v_{\parallel}$, $v_{\perp}$ are the velocity vector projections onto the axes x' and z', respectively. It is seen that for such a choice of the coordinate system, the set of equations (7.65) falls into two independent equations for the propagation of two types of perturbations for which the velocity vectors are perpendicular. With the field off, when $\eta_r = 0$, both equations (7.66) are identical. With the field on, effective viscosities for different perturbations, determined by relations (7.67), are different.

The solution to equations (7.66) will be sought as complex harmonics

$$v_{\parallel} = v_{\parallel *}\exp i\,(\kappa_{\parallel} z - \omega_{\parallel} t)\,;\ v_{\perp} = v_{\perp *}\exp i\,(\kappa_{\perp} z - \omega_{\perp} t) \qquad (7.68)$$

Here, $v_{\parallel *}$, $v_{\perp *}$ are complex amplitudes. The real and imaginary parts of the generalized wave number κ characterize, correspondingly, the length of the perturbations and their attenuation in space, while the real and imaginary parts of ω specify the frequency and attenuation of the perturbation in time. The conditions for a nontrivial solution to equations (7.66) are of the form

$$i\omega_{\parallel}\rho = \kappa_{\parallel}^2\eta_{\parallel},\ i\omega_{\perp}\rho = \kappa_{\perp}^2\eta_{\perp} \qquad (7.69)$$

Let us consider the main relationships which describe the distribution of viscous stresses in a fluid. In the geometry under consideration, the predominant contribution is made by the stress tensor components constituting the force $\vec{\sigma}'_y \equiv \vec{\sigma}$ acting upon a unit area normal to the direction of propagating perturbations. Substitute perturbations (7.68) successively into (6.47), (6.28) to obtain expressions for $\vec{\sigma}$ in various perturbations

$$\vec{\sigma}_{\parallel} = i\kappa_{\parallel}\,\eta_{\parallel}\,v_{\parallel}\,\vec{i}'_x\ ,\ \vec{\sigma}_{\perp} = i\kappa_{\perp}\,\eta_{\perp} v_{\perp}\,\vec{i}_{z'} \qquad (7.70)$$

Here $\vec{i}_{x'}$, $\vec{i}_{z'}$ are the unit vectors of the hatched coordinate system.

Two linearly independent solutions (7.68) whose characteristics satisfy dispersion equations (7.69) may be used to build the solutions to a series of boundary-value problems and initial-value problems. Here, we shall consider the problem on the generation of perturbations by a flat plate oscillating in an unlimited fluid. In this case it is necessary to try a solution to satisfy the boundary conditions

$$v_x = v_* e^{-i\omega t}, \; v_y = 0 \text{ at } y = 0; \tag{7.71}$$

$$v_x = 0 \qquad \text{at } y \to \infty. \tag{7.72}$$

Here, ω is a material quantity. From (7.69) we find the expressions for the wave numbers of the perturbations generated by the plate:

$$\kappa_{\parallel} = \pm (i+1)\sqrt{\frac{\rho\omega}{2\eta_{\parallel}}}, \qquad \kappa_{\perp} = \pm (i+1)\sqrt{\frac{\rho\omega}{2\eta_{\perp}}} \tag{7.73}$$

Perturbations propagating in opposite directions correspond to different signs of κ. We shall consider perturbations propagating in the positive direction of the axis z. In this case, the condition of perturbation attenuation far from the plate (7.72) will be satisfied by wave numbers (7.73) with the sign "+". Thus, the problem solution is a superposition of two waves of different types

$$\vec{v} = v_{\parallel}\vec{i}_{x'} + v_{\perp}\vec{i}_{z'} \tag{7.74}$$

Here, $v_{\parallel}$, $v_{\perp}$ are determined by relations (7.68). In order to satisfy the boundary conditions on the plate (7.71), the plate velocity vector must be resolved into two components in the hatched coordinate system. The component along the axis x' generates only the ∥-type wave while the z-component generates only the ⊥-type wave. Therefore, for the amplitude values of different types of perturbations we have

$$v_{\parallel *} = v_* \cos\delta = \frac{v_* e_x}{\sqrt{e_x^2 + e_z^2}}, \qquad v_{\perp *} = v_* \sin\delta = \frac{v_* e_z}{\sqrt{e_x^2 + e_z^2}} \tag{7.75}$$

Here, δ is the angle between the axes x' and z'. If the plate oscillates in the direction of the axis x' or z', then one wave of a corresponding type is generated. In these cases, the direction of the perturbation velocity vector is parallel to the plate oscillation in the entire space. In the general case, perturbations of both types are generated and, hence, the velocity vector has components both in the plate oscillation direction and normal to it. Project velocity vector (7.74) onto the axes x and z, with account of (7.68) and (7.75), to find for these components the relations

$$\left.\begin{aligned} v_x &= (\cos^2\delta\cdot e^{i\delta_{\|}y} + \sin^2\delta\cdot e^{i\delta_{\perp}z})\, v_*\, e^{-i\omega t}\;; \\ v_z &= \frac{1}{2}\sin^2\beta\,(-e^{i\delta_{\|}z} + e^{i\delta_{\perp}z})\, v_*\, e^{-i\omega t} \end{aligned}\right\} \qquad (7.76)$$

Distribution of viscous stresses in the fluid is the superposition of stresses in different wave

$$\vec{\sigma} = \vec{\sigma}_{\|} + \vec{\sigma}_{\perp} \qquad (7.77)$$

Here, $\vec{\sigma}_{\|}$, $\vec{\sigma}_{\perp}$ are determined by relations (7.70). Perform scalar multiplication of (7.77) by $\vec{i}_v$ and $\vec{i}_1$, to obtain the components of the tensor of viscous stresses which determine viscous friction forces on the plate

$$\sigma_{xy} = \frac{1-i}{\sqrt{2}}\sqrt{\rho\omega\eta_e}\, v_*\, e^{-i\omega t};\quad \sigma_{zy} = \sin 2\delta\,\frac{1-i}{2\sqrt{2}}\sqrt{\rho\omega}(\sqrt{\eta_{\|}} - \sqrt{\eta_{\perp}})\, v_*\, e^{-i\omega t} \qquad (7.78)$$

where $$\eta_e = ((\sqrt{\eta_{\|}}\, e_x^2 + \sqrt{\eta_{\perp}}\, e_z^2)/(e_x^2 + e_z^2))^2 \qquad (7.79)$$

Magnetoviscous Effect

A vibrating-plate viscometer is used to measure viscous stresses in the direction of plate oscillations. The longitudinal component of viscous stresses is presented in the form similar to viscous stresses in a normal fluid, with the only difference that the effective viscosity η_e acts as dynamic viscosity. Therefore, by processing the MF measurements according to the methods developed for a normal fluid we shall obtain effective viscosity related to the coefficients of dynamic and rotational viscosities and field orientation by (7.79). For the relative magnetoviscous effect, (7.79) yields the expression

$$S = \eta_e/\eta - 1 = (\sqrt{1+S}\,\cos^2\delta + \sqrt{1+Se_z^2}\,\sin^2\delta)^2 - 1 \qquad (7.80)$$

This relation differs from similar ones for the Couette (7.20) and Poiseuille (7.59) flows by that, in the general case, the magnetoviscous effect in the vibrating-plate viscometer is a nonlinear function of the parameter S. This is due to distortions of the velocity profile at applying the magnetic field. There are, however, some particular cases when relation (7.80) is linear in the parameter S. Thus, for plate oscillations in the directions of axes x' and y' we have, correspondingly,

$$S_{\|x'} = S,\; S_{\|y'} = Se_z^2 \qquad (7.81)$$

In diluted fluids, when the parameter S is small, relation (7.80) may be linearized in S at random orientations of the field. In this case

$$S_{\parallel} = S(1 - e_z^2) \tag{7.82}$$

Transverse Stresses

The second relation (7.78) represents transverse stresses which tend to change the direction of plate motion. From (7.78) obtain the expression for the dimensionless parameter of transverse stresses equal to the tangent of the angle between the total vector of viscous stresses and the plate direction

$$S_{\perp} = \frac{\sigma_{xy}}{\sigma_{zy}} = \frac{\sqrt{\eta_{\perp}} - \sqrt{\eta_r}}{2\sqrt{\eta_e}} \sin 2\delta = \frac{e_x e_z (\sqrt{1-S} - \sqrt{1+Se_y^2})}{\sqrt{1+S}\, e_z^2 + \sqrt{1+Se_y^2}\, e_x^2} \tag{7.83}$$

Determine the field orientation at which the parameter of transverse stresses is maximum. From the condition $\partial S_{\perp}/\partial e_z = 0$ find

$$e_{zm} = \sqrt{\frac{(1 - e_y^2)\sqrt{1+S}}{\sqrt{1+S} + \sqrt{1+Se_z^2}}} \tag{7.84}$$

At such orientation, obtain from (7.83)

$$S_{\perp m} = 0.5(\sqrt{1+S} - \sqrt{1+Se_y^2}) \tag{7.85}$$

It is seen that the absolute maximum of $S_{\perp}$ is attained when the field vector is in the plate plane, i.e. $e_y = 0$. Then from (7.85) obtain

$$S_{\perp *} = 0.5(\sqrt{1+S} - 1) \tag{7.86}$$

For diluted fluids

$$e_{x*} = (1 + S/4)\sqrt{2}/2; \; S_{\perp *} = S/4 \tag{7.85}$$

As follows from (7.82), at $e = e_*$ the relative magnetoviscous effect $S_{\parallel *} = S/2$ and, hence, $S_{\perp *} = S_{\parallel *}/2$. The maximum value of the dimensionless parameter of transverse stresses is half the magnetoviscous effect at constant field orientation.

CHAPTER

EIGHT

THE EFFECT OF TRANSVERSE STRESSES ON THE STRUCTURE OF TWO-DIMENSIONAL FLOWS IN A UNIFORM FIELD

The tensor of MF viscous stresses may be presented in the form

$$\sigma'_{ik} = \sigma'^{s}_{ik} + \sigma'^{a}_{ik} = 2\eta_r \, \epsilon_{ikl}\omega_l + \eta_{iklm} \cdot \frac{\partial v_l}{\partial x_m} \tag{8.1}$$

where $\vec{\omega} = e \times \dfrac{d\vec{e}}{dt}$ is the vector of the instantaneous angular rotational velocity of the field; η_{ikim} is the viscosity tensor for which, with account of (6.26), (6.34) and (6.36), obtain the expression

$$\eta_{iklm} = \eta(\delta_{im}\delta_{kl} + \delta_{il}\delta_{km}) + \eta_r(\delta_{il}e_k e_m - \delta_{im}e_l e_k - \\ - \delta_{kl}e_i e_m + \delta_{km}e_l e_i) \tag{8.2}$$

Thus, the rotational viscosity mechanism leads to that the total tensor of viscous stresses becomes an anisotropic function of the deformation rate tensor. The anisotropy induced by the field may stipulate an essential rearrangement of the structure of ferrohydrodynamic flows when the field is applied. Such a rearrangement may be exemplified by the generation of perturbations by the plate oscillating in its plane in the fluid normal to the plate direction, discussed in 7.6. On the other hand, as follows from sections 7.4 and 7.5, the application of a uniform field of random orientation does not change the structure of the Couette flow or the Poiseuille flow in an elliptical tube. The problem as to which of these phenomena is most widely encountered may be solved when studying flows more complex than one-dimensional flows considered in Chapter 7.

In the present chapter, this problem will be solved within the model of two-dimensional flows the characteristics of which are constant along some linear axis. In the Newtonian fluid there are two types of motion of such a structure. They are plane and axial flows with the velocity vectors, correspondingly normal and parallel to the axis. Because of the isotropic nature of the Newtonian tensor, viscous friction forces for plane motions lie in the same plane as the velocity vector; in case of axial flows, they are directed along the flow. Therefore, plane and axial flows may exist independently.

Transverse stresses may occur in MFs moving in a magnetic field as a result of the anisotropy of the viscous stress tensor. In a primary plane flow, they impel the fluid to move axially; in case of an axial flow, they may generate a plane component. Consequently, in the situations when transverse stresses are at play, it will be less possible that plane and axial flows may exist independently. Two-dimensional MF flows should therefore be regarded as three-dimensional ones, with the plane and axial motion components being interdependent. Now we shall dwell upon such motions.

8.1 TWO-DIMENSIONAL EQUATIONS

Let us transform equations (7.2) for two-dimensional flows, the characteristics of which are independent of one Cartesian coordinate. Assume that the velocity vector does not change in z direction, i.e., $\vec{v} = \vec{v}(x, y)$. Let us resolve the vectors and vector operators entering (7.2) into the components directed along or normal to the axis z: $\nabla = \partial \vec{i}_z / \partial_z + \nabla_\perp$; $\vec{e} = e_z \vec{i}_z + \vec{e}_\perp$; $\vec{v} = v_z \vec{i}_z + \vec{v}_\perp$; $\rho\vec{g} = \rho g_z \vec{i}_z + \rho \vec{g}_\perp$. Substituting these vector resolutions into (7.2), we get the set of equations

$$\left.\begin{aligned} &\rho\, d v_z/dt = -\partial p/\partial z + (\eta + \eta_r)\, \nabla_\perp^2 v_z + \eta_r \vec{i}_z \cdot \nabla_\perp \times \vec{e}_\perp (\vec{e} \cdot \nabla_\perp \cdot \vec{v}) + \\ &+ \rho g_z\, ; \\ &\rho d\vec{v}_\perp/dt = -\nabla_\perp p + (\eta + \eta_r) \nabla^2 v_\perp + \eta_r\, \nabla \times \vec{i}_z (\vec{e} \cdot \nabla_\perp \times \vec{v}) + \rho \vec{g}_\perp \end{aligned}\right\} \quad (8.3)$$

Here, $d/dt = \partial/\partial t + v_\perp \nabla_\perp$; $\nabla_\perp$ is a two-dimensional nabla. System (8.3) supplements the continuity equation which in the geometry under consideration is of the form $\nabla_\perp \cdot \vec{v}_\perp = 0$.

Obtain restrictions imposed by the two-dimensional flow condition on pressure and volume force distribution $\rho\vec{g}_\perp$. To do this, eliminate pressure from equation (7.2) by taking its operation $\nabla \times$. As the velocity vector does not change along the axis z, then $\nabla \times \rho\vec{g}$ should also be independent of z. This restriction is satisfied by the following distribution of the volume force

$$\rho\vec{g} = \rho_\perp(x,y,t)\vec{g}_\perp + [\rho_\perp(x,y,t) + \rho_z(z,t)]g_z\vec{i}_z \tag{8.4}$$

As follows from the first equation (8.3), the two-dimensional velocity profile condition imposes restrictions on the pressure dependence on the coordinate z:

$$-\partial p/\partial t + \rho_z g_z = -c = \text{const} \tag{8.5}$$

Using the known formulae of vector analysis, in equations (8.3) transform all the terms including the vector $\vec{e}$:

$$2\vec{i}_z\cdot\nabla\times(\vec{e}\cdot\vec{\Omega}) = (e_\perp\cdot\nabla)^2 v_z - e_\perp\nabla^2 v_z - 2e_z(n\cdot\nabla)\Omega_z \tag{8.6}$$

$$2\nabla\times\vec{i}_z(\vec{e}\cdot\vec{\Omega}) = -e_z^2\nabla^2\vec{v}_\perp + e_z[\vec{e}_\perp\nabla^2 v_z - \nabla(\vec{e}_\perp\cdot\nabla)\,v_z] \tag{8.7}$$

Here and below, the notations $\nabla \equiv \nabla_\perp$, $2\vec{\Omega} = \nabla\times\vec{v}$, $\vec{n} = \vec{i}_z \times \vec{e}$ are used. Relations (8.5) through (8.7) make it possible to present set (8.3) in the form

$$\rho\, d\vec{v}_\perp/dt = -\nabla_\perp[p+\eta_r e_z(\vec{e}_\perp\cdot\nabla)v_z] + (\eta+\eta_r e_\perp^2)\nabla^2\vec{v}_\perp + \eta_r e_z(\vec{n}\cdot\nabla)\nabla^2 v_z +$$
$$+\rho_\perp g_\perp \tag{8.8}$$

$$\rho\, d v_z/dt = -c + (\eta+\eta_r e_z^2)\nabla^2 v_z + \eta_r(\vec{e}_\perp\cdot\nabla)^2 v_z - 2\eta_r(\vec{n}\cdot\nabla)\Omega_z + \rho_\perp g_z \tag{8.9}$$

Taking the operation $\nabla\times$ from equation (8.8) and projecting the equation thus obtained onto the axis z, we find, with regard to the continuity equation,

$$2\rho d\Omega_z/dt = 2(\eta + \eta_r e_\perp^2)\nabla^2\Omega_z - \eta_r e_z(\vec{n}\cdot\nabla)\nabla^2 v_z + (\nabla\rho_\perp \times \vec{g}_\perp)_z \tag{8.10}$$

Relations (8.9) and (8.10) constitute a set of equations for z-components of velocity and velocity curl. With the magnetic field off, when $\eta_r \to 0$, the set will be split. Equation (8.10) for the axial component of the velocity curl describes plane flows, two velocity components of which are in the plane normal to the axis z. It may be integrated irrespective of (8.9). Equation (8.9) describes axial motions related to plane ones via the convective term $d/dt = \partial/\partial t + \vec{v}_\perp\cdot\nabla$. At $\eta_r \neq 0$, the equations for plane and axial flows are interrelated. Hence follows the main qualitative feature of the field effect on the class of two-dimensional flows under consideration. With a uniform stationary field on, the primary axial flow may generate a plane motion component. Vice versa, the primary plane flow may become a source of an axial flow.

Proceeding from set (8.9), (8.10), we can obtain general conditions for splitting two-dimensional flows. Assuming in equation (8.9) that the source dependent on the plane flow component is equal to zero, obtain the conditions under which the plane source does not generate by an axial flow

$$\eta_r e_z(\vec{i}_z \times \vec{e} \cdot \nabla)(\nabla \times \vec{v})_z = 0 \tag{8.11}$$

At $\eta_r \neq 0$, this condition is satisfied for the simplest orientations of the field: when it is either parallel or normal to the axis z. At random orientation ($e_z e_y \neq 0$), (8.11) holds for the flows for which z, the vorticity component, changes only in the $\vec{e}_\perp$ direction. This condition is satisfied, for example, by the Couette profile for which vorticity is constant throughout the flow.

Let us elucidate the conditions when the axial flow does give rise to the plane one. By setting the source in equation (8.10) to zero, which is dependent on the axial flow, obtain

$$\eta_r e_z(\vec{i}_z \times \vec{e})\nabla)\nabla^2 v_z = 0 \tag{8.12}$$

In addition to the situation when $e_z e_\perp = 0$, this equality holds for the flows for which the Laplacian of velocity distribution changes only in the vector $\vec{e}_\perp$ direction. The flow obeying condition (8.12) may be exemplified by the flow due to a pressure drop in the elliptical channel for which the Laplacian is constant.

Let us transform set (8.9), (8.10) to be appropriate for further application. The continuity equation $\nabla \cdot \vec{v}_\perp = 0$ allows $\vec{v}_\perp$ to be expressed via the stream function ψ:

$$\vec{v}_\perp = \nabla \times \psi \vec{i}_z = \nabla\psi \times \vec{i}_z \tag{8.13}$$

For z-component of the velocity curl, it gives the relation

$$2\Omega_z = -\nabla^2\psi \tag{8.14}$$

Eliminating $\vec{v}_\perp$ from set (8.9), (8.10), using (8.13) and (8.14), obtain a closed set for the axial velocity component and the stream function

$$\rho dv_z/dt = -c + (\eta + \eta_r e_z^2)\nabla^2 v_z + \eta_r (\vec{e}_\perp \cdot \nabla)^2 v_z + \eta_r e_z (\vec{n} \cdot \nabla)\nabla^2 \psi + \rho_\perp g_z \tag{8.15}$$

$$\rho d(\nabla^2 \psi)/dt\, (\nabla^2 \psi)/dt\, (\eta + \eta_r e_\perp^2)\nabla^4 \psi + \eta_r e_z (\vec{n} \cdot \nabla)\nabla^2 v - (\nabla \rho_\perp \times \vec{g}_\perp)_z$$

where

$$d/dt = \partial/\partial t + (\nabla \psi \times \vec{i}_z \cdot \nabla) \tag{8.16}$$

Equations of Two-Dimensional Flows in a Quickly Rotating Field

The set of equations (8.15), (8.16) is valid both for a stationary and variable field. In case of a rotating field, the components of the unit vector $\vec{e}$ are pulsating values. However, if the field rotation frequency ω essentially exceeds the natural frequencies of hydrodynamic flow pulsations, it is possible to derive a set of equations with time-constant coefficients. In this case, it should be regarded that for a time interval equal to the period of field rotation the velocity distribution does not practically vary, which admits an independent time averaging of the unit vector $\vec{e}$ components in (8.15), (8.16). Taking account of (7.5) and (7.11), find

$$\left.\begin{aligned} (e_z^2)_{\mathrm{m}} &= 0.5e_{0z}^2\ ; \quad (e_\perp^2)_{\mathrm{m}} = 1 - 0.5e_{0z}^2\ ; \\ (\vec{e}_\perp\cdot\nabla)_{\mathrm{m}}^2 &= 0.5(\vec{e}_{0\perp}\cdot\nabla)^2 + 0.5(\vec{e}_1\cdot\nabla)^2; \\ (e_z(\vec{n}\cdot\nabla)_{\mathrm{m}} &= 0.5(e_{0z}(\vec{i}_z \times e_0\cdot\nabla) = 0.5e_{0z}(\vec{n}_0\cdot\nabla) \end{aligned}\right\} \qquad (8.17)$$

The last operator in (8.17) is responsible for interdependence between the plane and axial flow components. In the rotating field, unit vector $\vec{e}$ plays the part similar to that played by unit vector $\vec{e}_0$. Axial and plane flows split at $e_{0z}e_{0\perp} = 0$. This condition is satisfied if the vector of angular velocity is directed along or normal to the axis z. In the opposite case, two-dimensional flows are three-dimensional ones.

Equations of Two-Dimensional Perturbations of Plane-Parallel Flows

In what follows the stability of MF shear flows will be discussed. Let us formulate a set of equations that describes the dynamics of small two-dimensional perturbations against the background of a plane-parallel fluid flow in a uniform field. Let there be an unperturbed flow for which velocity and pressure distributions are $\vec{v}_0$ and p_0. Present the perturbed velocity and pressure fields as superpositions: $\vec{v}_0 + \vec{v}$, $p_0 + p$. Here, $\vec{v}$ and p are small perturbations of the main stream.

Let the main stream direction be specified by the unit vector $\vec{i}$ and velocity distribution be only dependent on the coordinate y, i.e. $\vec{v}_0 = v_0(y)\,\vec{i}$. In a uniform field whose orientation is free from restrictions, at least two flows of such a structure may be realized. It is a simple shear flow discussed in 7.3 and the pressure drop-assisted flow in the plane-parallel channel considered in 7.5. We specify that the perturbation characteristics are independent of the coordinate z. As the unperturbed flow characteristics do not depend on z, either, the resultant flow is described by set (8.9), (8.10) and set (8.15), (8.16) stemming from the latter one. So, the behavior of small two-dimensional perturbations in plane-parallel flows, is described by set (8.9), (8.10) where nonlinear convective terms should be linearized by small perturbations. In this case they have the form

$$\rho dv_z/dt = \rho(\partial v_z/\partial t + v_{0x}\,\partial v_z/\partial x + v_y\,\partial v_{0z}/\partial y) \qquad (8.18)$$

$$2\rho d\Omega_z/dt = 2\rho(\partial\,\Omega_z/\partial t + v_{0x}\partial\Omega_x/\partial x + v_y\partial\Omega_{0z}/\partial y) \qquad (8.19)$$

The set of equations for small perturbations is essentially simplified when the velocity vector of the main stream is directed along the axis z; in other words, when we consider the behavior of small perturbations whose characteristics are constant along the unperturbed flow. In this case we obtain from (8.15), (8.16) with regard to (8.13), (8.14) and (8.19):

$$\rho\left(\frac{\partial v_z}{\partial t} - \frac{\partial v_{0z}}{\partial y}\frac{\partial \psi}{\partial x}\right) = (\eta + \eta_r e_z^2)\nabla^2 v_z + \eta_r (\vec{e}_\perp \cdot \nabla)^2 v_z + \eta_r e_z (\vec{n}\cdot\nabla)\nabla^2 \psi \tag{8.20}$$

$$\rho\partial(\nabla^2\psi)/\partial t = (\eta + \eta_r e_\perp^2)\nabla^4\psi + \eta_r e_z (\vec{n}\cdot\nabla)\nabla^2 v_z \tag{8.21}$$

The velocity components normal to the main stream are estimated in terms of the stream function in accordance with the relations:

$$v_x = \partial\psi/\partial_y, \; v_y = -\partial\psi/\partial x \tag{8.22}$$

which follow from (8.13).

With the magnetic field off, the plane-parallel flows are stable with respect to perturbations whose characteristics do not change downstream of the flow at arbitrary flow velocities, i.e., at any Reynolds numbers. This conclusion follows from (8.20) and (8.21) where it is necessary to set $\eta_r = 0$. With the field on, the plane and axial perturbation components are interrelated. In the axial flow equation there is a source related to the main stream. As will be shown later, this source may provide the energy supply required for the perturbations to develop. Thus, set (8.20), (8.21) is the simplest model which demonstrates the specificity of the field effect on hydrodynamic stability of MFs since the flow structure under consideration excludes the instability mechanism of the Orr—Sommerfeld type known in the dynamics of a normal viscous fluid.

Boundary Conditions

Let us formulate the boundary conditions required for integration of set (8.20), (8.21). If the flow region is bounded with a solid surface, then the conditions of fluid adhesion to the boundary are satisfied. For a fixed plane boundary (normal to the axis y) these conditions have the form

$$v_z = v_x = v_y = 0 \tag{8.23}$$

Hence, with allowance for relations (8.22), follow the conditions for the stream function

$$\psi = \partial\psi/\partial y = 0 \tag{8.24}$$

The condition of vanishing viscous stress vector $\vec{\sigma}'_y = 0$ holds on a free boundary. The projections of this equality onto the coordinate axis give the relations

$$\left.\begin{aligned} \sigma_{xy} &= \eta\partial v_x/\partial y + \eta_r\left[(1-e_z^2)\partial v_x/\partial y + e_z e_x\,\partial v_z/\partial y - e_z e_y \partial v_z/\partial x\right] = 0; \\ \sigma_{zy} &= \eta\partial v_z/\partial y + \eta_r\left[(1-e_z^2)\,\partial v_z/\partial y + e_x e_y\,\partial v_z/\partial x + e_x e_z\,\partial v_x/\partial y\right] = 0 \end{aligned}\right\} \tag{8.25}$$

Besides, the condition of fluid adhesion is still valid

$$v_y = 0 \tag{8.26}$$

A linear combination of conditions (8.25) at $e_x e_z \neq 0$ may be obtained by the relation

$$e_z\partial v_z/\partial_y + e_x\partial v_x/\partial_y = 0 \tag{8.27}$$

By the use of this relation, condition (8.25) may be presented in a simpler form

$$(\eta + \eta_r\, e_y^2)\,\partial v_x/\partial y - \eta_r\, e_z e_y\,\partial v_z/\partial x = 0 \tag{8.28}$$

$$(\eta + \eta_r\, e_y^2)\;\partial v_z/\partial y + \eta_r\, e_x e_y\,\partial v_z/\partial x = 0 \tag{8.29}$$

The system of boundary conditions (8.26), (8.28) and (8.29) contains all the three velocity components. Using the continuity equation $\partial v_x/\partial x + \partial v_y/\partial y = 0$, the velocity component x may be excluded from equation (8.28) to obtain

$$(\eta + \eta_r e_y^2)\partial^2 v_y/\partial y^2 + \eta_r e_r e_z \partial^2 v_z/\partial x^2 = 0 \tag{8.30}$$

Thus, the alternative formulation of free boundary conditions is a set of equations (8.26), (8.29) and (8.30). With the magnetic field off, when $\eta_r = 0$, it results in the boundary conditions used in viscous fluid dynamics: for axial flows

$$\partial v_z/\partial y = 0 \tag{8.31}$$

for plane flows

$$v_y = \partial^2 v_y/\partial y^2 = 0 \tag{8.32}$$

8.2 SIMPLE SHEAR INSTABILITY

Let there be a steady fluid flow in a state of simple shear between the planes $2l$ apart. Direct the axis z of the Cartesian coordinates downstream. In this case, velocity distribution may be presented as

$$\vec{v}_0 = 2\Omega/_0 y \vec{i}_z \tag{8.33}$$

Here, Ω_0 is the velocity curl projection onto the axis x. Let us study the development of small perturbations whose characteristics along the axis z are constant. The dynamics of such perturbations is described by set (8.20), (8.21) where $v_0 = 2\Omega_0 y$. Consider normal perturbations that are periodic along the axis x:

$$v_z = v_a(y)\exp(i\kappa x - \sigma t),\ \psi = \psi_a(y)\exp(i\kappa x - \sigma t) \tag{8.34}$$

Estimate the amplitude of perturbations v_a and ψ_a, using a set of ordinary differential equations that follows from (8.20) and (8.21):

$$-\sigma v_a - 2i\Omega_0\psi_a = (\nu - \nu_r e_z^2)\nabla^2 v_a + \nu_r (\vec{e}_\perp \cdot \nabla)^2 v_a + \nu_r e_z (\vec{n}\cdot\nabla)\nabla^2\psi_a \tag{8.35}$$

$$-\sigma\nabla^2\psi_a = (\nu+\nu_r e_\perp^2)\nabla^4\psi_a + \nu_r e_z (\vec{n}\cdot\nabla)\nabla^2 v_a \tag{8.36}$$

Here, $\nabla^2 = d^2/dy^2 - \kappa^2,\ \vec{n}\cdot\nabla = e_x d/dy - i\kappa e_y$;

$$(\vec{e}_\perp\cdot\nabla)^2 = e_y^2 d^2/dy^2 + 2i\kappa e_x e_y d/dy - e_x^2\kappa^2,\quad \nu_r = \eta_r/\rho,\ \nu = \eta/\rho \tag{8.37}$$

The boundary conditions for amplitudes at $y = \pm d$ have the form: on rigid boundaries

$$v_a = \psi_a = d\psi_a/dy = 0 \tag{8.38}$$

on free boundaries

$$\left.\begin{aligned} &(\nu + \nu_r e_y^2)\, d^2\psi_a/dy^2 - \nu_r e_z e_y i\kappa v_a = 0; \\ &(\nu + \nu_r e_y^2)\, dv_a/dy + \nu_r e_x e_y i\kappa v_a = 0,\ \psi_a = 0 \end{aligned}\right\} \tag{8.39}$$

The set of equations (8.35), (8.36) subject to boundary conditions (8.38) or (8.39), is a linear uniform boundary-value problem. It has a nontrivial solution at certain values of the frequency σ. These values are eigenvalues of the problem; their corresponding solutions are eigenfunctions. Let us exclude v_a from set (8.35), (8.36). This results in the equation

$$\{[\sigma + \nu\nabla^2 + \nu_r (\vec{e}_\perp\cdot\nabla)^2][\sigma + (\nu+\nu_r)\nabla^2] - 2i\kappa\Omega_0\nu_r e_z (\vec{n}\cdot\nabla)\}\nabla^2\psi = 0 \tag{8.40}$$

A similar equation describes the v_a distribution.

Consider a model case that allows a simple analytical investigation. Assume that the field vector is in the plane $[y,z]$, i.e. $e_x = 0$. In this case, as follows from (8.37), (8.35) and (8.36), we must set

$$(\vec{n} \cdot \nabla) = -i\kappa e_y; \; (\vec{e}_\perp \cdot \nabla)^2 = e_y^2 d^2/dy^2 \tag{8.41}$$

No axial perturbation component on the layer boundaries

$$v_a = 0 \tag{8.42}$$

fluid adhesion to the boundaries

$$\psi_a = 0 \tag{8.43}$$

and no transverse stresses on the boundaries (the first equation) from (8.25) which, with regard to (8.38), is reduced to the condition

$$\partial^2 \psi_a / \partial y^2 = 0 \tag{8.44}$$

are used as boundary conditions. The combination of boundary conditions (8.42) through (8.44), seems to be rather artificial, since both condition (8.42), characteristic of a rigid boundary, and condition (8.44), of a free boundary, must hold in one and the same plane. However, system of boundary conditions (8.42) through (8.44) may not change the qualitative meaning of the problem as no energy is supplied to the perturbation via the boundary. Indeed, the work done by external forces to offset the viscous friction forces, generated by the main stream perturbations, comes to naught, for at $y = \pm d$

$$-\vec{\sigma}_y \cdot \vec{v} = -\sigma_{xy} v_x - \sigma_{zy} v_z = 0$$

due to the fulfilment of (8.42) and the first equation from (8.25). Therefore, the discussion of the formulated problem reveals the internal mechanisms which may destroy a simple shear flow. Its solution is elementary. For odd and even perturbations, correspondingly, we have

$$\left.\begin{aligned} v_a &= v_* \cos \frac{\pi y}{2d}(n+1); \; \psi_a = \psi_* \cos \frac{\pi y}{2d}(n+1), n = 0, 2, 4, ... \\ v_a &= v_* \sin \frac{\pi y}{2d}(n+1); \; \psi_a = \psi_* \sin \frac{\pi y}{2d}(n+1), n = 1, 3, ... \end{aligned}\right\} \tag{8.45}$$

Substitution of (8.46) and (8.47) into (8.40) results in a dispersion equation specifying the spectrum of decrements

$$[\sigma - (\nu + \nu_r)(\kappa^2 + \delta^2)][\sigma - \nu(\kappa^2 + \delta^2) - \nu_r e_y^2 \delta^2] = 2\Omega_0 \nu_r e_z e_y \kappa^2 \quad (8.46)$$

where $\delta = \pi(n + 1)/2d$, $n = 0, 1, 2. \ldots$ Henceforth, find the expression for perturbation damping decrements:

$$2\sigma_{1,2} = 2\nu(\kappa^2+\delta^2) + \nu_r(\kappa^2+\delta^2+\delta^2 e_y^2) \pm \sqrt{\nu_r^2(\kappa^2+\delta^2-\delta^2 e_y^2)^2 + 8\Omega_0 \nu_r e_z e_y \kappa^2} \quad (8.47)$$

The qualitative picture of σ vs. wave number κ is shown in Fig. 8.1.

The dashed curves represent the damping decrements of two independent types of perturbations realized without a shear flow, i.e., at $\Omega_0 = 0$. The equilibrium perturbations are damped monotonically as their frequencies are real and positive. The solid curves in the sector between the dashed lines represent the damping frequencies when the factor $\Omega_0 e_z e_y < 0$. The solid curves in the external region correspond to $\Omega_0 e_z e_y > 0$. As is seen, at $\Omega_0 e_z e_y < 0$, the real part of the frequencies is positive, which points to the damping of perturbations. The constant shear, however, imparts specific qualitative feature to the frequency spectrum. When κ is fixed, there are two different real modes only in long- and short-wavelength regions. In the intermediate region, expression (8.49) yields two conjugate complex roots, which testifies to the presence of oscillating perturbations.

At $\Omega_0 e_z e_y > 0$, equation (8.48) has two real roots. One of them, however, may reverse its sign, thus pointing to the existence of developing perturbations. Set $\sigma = 0$ in (8.48) to obtain the neutral curve equation both in dimensional and dimensionless form

$$\left. \begin{aligned} 2\Omega_0 &= \frac{(\nu + \nu_r)(\kappa^2+\delta^2)[\nu(\kappa^2+\delta^2) + \nu_r e_y^2 \delta^2]}{\nu_r e_z e_y \kappa^2} \\ R &= \frac{(1+S)(\hat{\kappa}^2 + m\pi^2)[\hat{\kappa}^2 + m^2\pi^2(1+Se_y^2)]}{Se_z e_y \kappa^2} \end{aligned} \right\} \quad (8.48)$$

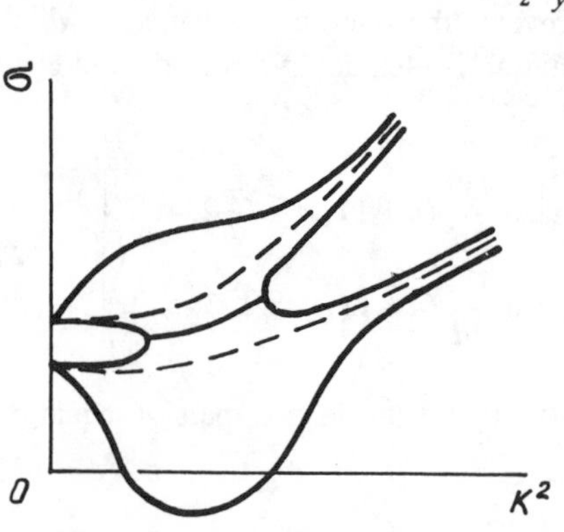

FIG. 8.1

Here, $R = 8\Omega_0 l^2/\nu$ is the Reynolds number for an unperturbed flow; $\hat{\kappa} = 2\kappa l$, $m = 1, 2, 3. \ldots$ Let us find a minimal Reynolds number at which non-damping perturbations may appear in the flow. As is seen from (8.48), most detrimental is the perturbation with even distribution v_a, ψ_a for which $m = 1$. From the condition $dR/dk = 0$ find that R has a minimum at

$$2\kappa_* = \pi\sqrt[4]{1 + Se_y^2} \tag{8.49}$$

Substitute (8.49) into (8.48) to find the critical Reynolds number

$$R_* = \frac{\pi^2}{Se_z e_y}(1 + S)(1 + \sqrt{1 + Se_y^2})^2 \tag{8.50}$$

As follows from (8.50), instability takes place only at $S \neq 0$. With the field off, when $S \to 0$, the critical Reynolds number goes into infinity.

Let us explain the sign of the product $\Omega_0 e_z e_y$ being a determinant factor of instability. For the case in Fig. 8.2a, the factor $\Omega_0 e_z e_y$ is positive while it is negative for the case shown in Fig. 8.2b. It can easily be seen that a simple shear can be unsteady if the velocity profile and the field direction pass through the same quarters of the coordinate system; otherwise, the flow is steady.

The analysis so performed allows certain conclusions to be made that the flow of fluid between rotating cylinders is steady. Fig. 8.2b shows the flow in a narrow gap in the field normal to the cylinder axes. For an arbitrary point in the gap $\Omega_0 e_z e_y = 0.5\Omega_0 \sin 2\varphi$. The angle φ is calculated from the field direction. Henceforth it follows that the regions for which the angle φ is close to $\pi/4$ and $(-3/4)\pi$ are most detrimental for developing instability. In these regions, the factor is positive and achieves maximum.

Short-Wave Approximation

Let us consider perturbations whose wavelength is much shorter than the layer width $2l$. The characteristics of such perturbations slowly vary in the Y-direction and may be simulated by perturbations dependent on one space variable. Instability parameters for such perturbations may be obtained through an exact solu-

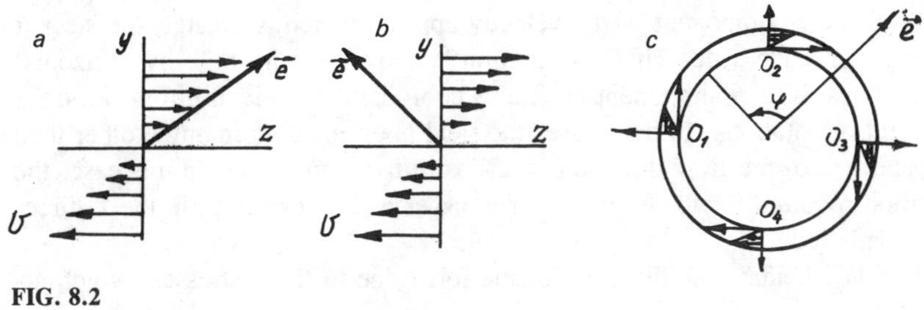

FIG. 8.2

tion of equation (7.2) in an arbitrary oriented field. Assume in (7.9) $v_z(y, t) = 2\Omega_0 y + v_z(x,t)$, $v_y = v_y(x,t)$ to obtain the set of equations

$$\left.\begin{aligned} \partial v_y/\partial t &= [\eta + \eta_r (1-e_z^2)]\partial^2 v_y/\partial x^2 + \eta_r e_y e_z \partial^2 v_z/\partial x^2; \\ \partial v_z/\partial t + 2\Omega_0 v_y &= [\eta + \eta_r (1-e_y^2)]\partial^2 v_z/\partial x^2 + \eta_r e_y e_z \partial^2 v_y/\partial x^2 \end{aligned}\right\} \quad (8.51)$$

This set allows a solution of the form

$$v_y = v_{xy}\exp(i\kappa x - \sigma t),\ v_z = v_{*z}\exp(i\kappa x - \sigma t)$$

provided κ and σ satisfy the characteristic equation

$$(\eta_{\|}\kappa^2 - \sigma\rho)(\eta_{\perp}\kappa^2 - \sigma\rho) = 2\Omega_0\rho\eta_r e_y e_z \quad (8.52)$$

Here,

$$\eta_{\|} = \eta + \eta_r;\ \eta_{\perp} = \eta + \eta_r e_x^2$$

With no steady shear ($\Omega_0 = 0$), when the coordinate system axes are matched, (8.52) becomes (7.69) and (7.67) describing the propagation of one-dimensional perturbations in the field at rest.

Equation (8.52) allows for the existence of a steady flow which is periodic in the X-direction. Assuming $\sigma = 0$ to find the relation between the shear flow rate of an unperturbed flow and the wave number of neutral perturbation

$$2\Omega_0 = \frac{\eta_{\|}\,\eta_{\perp}\,\kappa^2}{\rho\eta_r e_y e_z} = \frac{(\nu + \nu_r)(\nu + \nu_r e_x^2)\kappa^2}{\nu_r e_y e_z} \quad (8.53)$$

At $e_x = 0$, (8.53) is a short-wave asympote ($\kappa \gg \delta$) of expression (8.48).

Velocity distribution for one-dimensional perturbations is very simple to understand and provides a graphic notion of the instability mechanism. Solid horizontal lines in Fig. 8.3a show a section of the velocity profile, v_{0z}, of an unperturbed flow; solid vertical lines show the distribution of the velocity y-component of the perturbation localized on some segment of the axis x. As a results of the "stall" of the main stream by transverse perturbation, the perturbation of the z-component of the velocity appears which is greater, the steeper the unperturbed profile. This perturbation is shown in Fig. 8.3a by horizontal dashed lines. The z-component profile vs coordinate x is given in Fig. 8.3c for two field orientations. In both cases the field is oriented at an angle other than zero and $\pi/2$ to the flow direction, and lies in the plane $[x,y]$. In this case, the condition required for the existence of transverse stresses acting in the X-direction is satisfied.

Let us elucidate the direction of the force due to these stresses, which af-

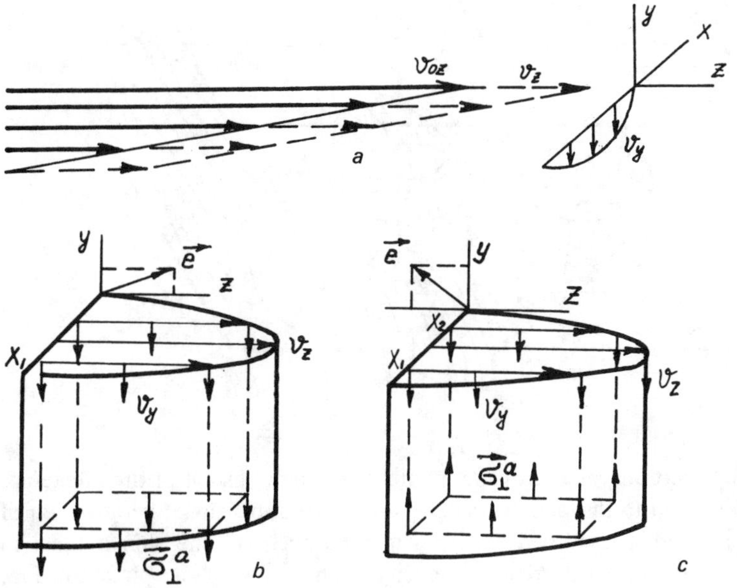

FIG. 8.3

fects the fluid layer contained between the planes $x = x_1$ and $x = x_2$. The mechanism of transverse stresses was discussed in detail in 7.3. It may easily be seen from Fig. 7.7 that, at $\Omega_0 e_y e_z > 0$, the force of transverse stresses, applied to both planes bounding the layer, has the same direction as the primary layer displacement specified by the vector v_y (cf. Fig. 8.3b). This case provides all the necessary grounds to speak about developing perturbations and instability onset. At $\Omega e_y e_z < 0$ (cf. Fig. 8.3c), the displacement of the fluid layer and the force of transverse stresses due to this displacement, are opposite in direction, which leads to the damping of perturbations. Similar processes take place in case of primary displacement of the layer in the direction opposite to the axis y.

8.3 INSTABILITY OF A PLANE-PARALLEL POISEUILLE FLOW

In MF flows with an even velocity profile, transverse stresses may be responsible for the instability mechanism somewhat different from the one of the simple shear discussed in 8.2. Let us interpret this mechanism for a plane-parallel Poiseuille flow presented schematically in Fig. 8.4. The unperturbed flow is in Z-direction. If the field vector is inclined to the axial flow direction ($e_z e_x \neq 0$), transverse stresses $\sigma_{\perp 0}$ will occur in the flow in the X-direction. Because of the linearity of the transverse stress profile, the fluid layers are affected by a volume force in the X-direction, whose density $F_{0\perp}$ is constant. It is

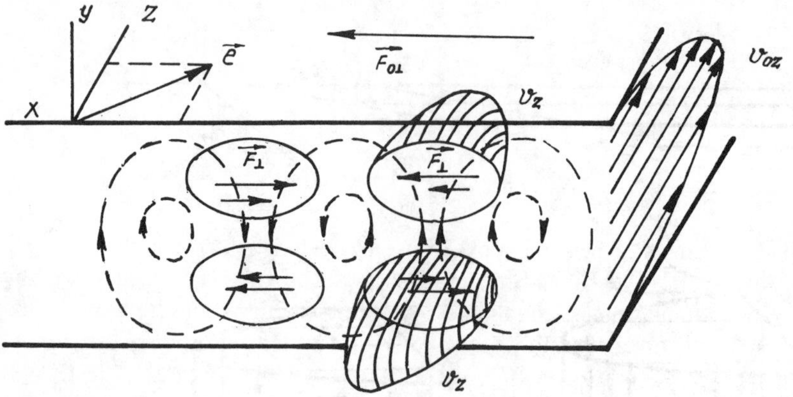

FIG. 8.4

balanced by the transverse pressure gradient. At some instant of time there appears, in addition to the main stream, a small two-dimensional perturbation of the stream function disturbed evenly in the Y-direction. The stream function isolines for transverse perturbation velocity components v_{0z} are plotted in Fig. 8.4 as hatched closed lines. Because of the "stall" of the unperturbed profile of the y-component of velocity perturbation, a perturbation with odd y-distribution of the z-component of velocity is generated. This perturbation is the greater, the more intensive is the unperturbed flow. The isolines of the z-component of velocity perturbation are presented in Fig. 8.4 as solid circular lines. If the directions of z-components of the main stream and perturbation coincide, then the directions of the volume transverse forces $\vec{F}_\perp$ and $\vec{F}_{\perp 0}$ they generate coincide as well. Vice versa, if a perturbation retards the main stream $F_\perp$, is opposite in direction to $F_{\perp 0}$. The $F_\perp$ field is an eddy and cannot be balanced by hydrostatic pressure. It is seen from Fig. 8.4 that the curl of the transverse velocity component and that of the transverse force field generated by it are displaced along the axis x by a quarter of the perturbation wavelength. As a result, at different points in the fluid, the force of transverse stresses may both maintain and damp the initial perturbation. The inertia velocity lag behind the stresses encountered in unsteady processes, is the mechanism that is able to coordinate the velocity and stress fields to enhance perturbations throughout the fluid. It should, therefore, be expected that the considered mechanism may stipulate vibrational instability. We shall now elucidate the conditions for such a coordination through numerical solution of the following problem.

Let us study the development of perturbations against the background of a plane-parallel flow caused by a pressure drop. The flow in the channel $2l$ in width is maintained by the pressure increment Δp on the section l along the channel axis. This flow was covered in 7.5. From relations (7.54) through (7.56) at $b/a = 0$ we have for unperturbed velocity and pressure distribution the identities

$$v_0 = \frac{1}{2\eta_e}\frac{\Delta p}{l}(y^2 - l^2) = v_{0*}(1 - \frac{y^2}{l^2});$$
$$p_0 = \frac{\Delta p}{l}(z + \frac{\eta_r}{\eta_e} e_z e_x) + A \qquad (8.54)$$

where

$$\eta_e = \eta + \eta_r\,(1 - e_x^2), A = \text{const}$$

The dynamics of perturbations, whose characteristics do not change along the main stream, is described by set (8.20), (8.21) where v_0 is determined by (8.54). For dimensionless amplitudes of normal perturbations periodic in the X-direction, we obtain from (8.35) and (8.36) a system of ordinary differential equations:

$$\left.\begin{aligned} &-\sigma v - i\kappa R y \psi = (1 + Se_z^2)\nabla^2 v + S(e_y^2 v'' - \kappa^2 e_x^2 v) + Se_z(e_x \nabla^2 \psi' - \\ &- i\kappa e_y \nabla^2 \psi); \\ &-\sigma \nabla^2 \psi = [1 + S(1 - e_z^2)]\nabla^4 \psi + Se_z(e_x \nabla^2 v' - i\kappa e_y \nabla^2 v) \end{aligned}\right\} \qquad (8.55)$$

Here,

$$\nabla^2 = \frac{d^2}{dy^2} - \kappa^2;\; S = \frac{\eta_r}{\eta};\; R = \frac{\rho d^3 \Delta p}{\eta_e \eta} = -\frac{\rho v_{0*} 2l}{\eta} \qquad (8.56)$$

κ, σ are a dimensionless wave number and a decrement, respectively. Chosen as characteristic quantities are: the layer half-width l for distance; $v_* = v/l$ for velocity and $t_* = l^2/v$ for time. We shall now supplement set (8.55) with boundary conditions (8.38) for rigid surfaces at $y = \pm 1$.

$$v = \psi = \psi' = 0 \qquad (8.57)$$

As follows from (8.54), in the main stream there is a pressure drop in the direction perpendicular to the channel axis. This pressure drop balances the transverse component of the volume force due to the interaction of the moving fluid and the magnetic field. Let us discuss the stability of this dynamic equilibrium. For the transverse pressure to appear, simultaneous existence of z- and x-components of the field in the layer plane is required. The field component normal to the layer plane is not a parameter to determine a transverse pressure drop. Therefore, we set $e_y = 0$ for convenience.

We shall solve boundary-value problems (8.55) through (8.57) for the eigenvalues using the Galerkin method. As a set of basic functions we shall choose the spectrum of two-dimensional perturbations of Newtonian fluid equilibrium. Assume in (8.55) $R = S = 0$ to obtain the equation

$$\nabla^2 v^{(0)} = -a v^{(0)};\; \nabla^4 \psi^{(0)} = -b\psi^{(0)} \qquad (8.58)$$

It is seen that the spectrum of two-dimensional viscous fluid perturbations falls into two independent groups. The first of equations (8.58) describes axial perturbations whose damping coefficients are a. The second equation specifies plane perturbations with coefficients b. The rest of eigenfunctions of equations (8.58), satisfying the boundary conditions on a rigid boundary, is a basis widely used for an approximate solution of two-dimensional problems pertaining to viscous fluid dynamics. The normalized eigenfunctions satisfy the following orthogonal conditions

$$\int_{-1}^{1} v_j^{(0)} v_e^{(0)}\, dy = \delta_{jl}\ , \quad \int_{-1}^{1} \psi_j^{(0)} \nabla^2 \psi_l^{(0)}\, dy = -\delta_{jl} \tag{8.59}$$

The cross relation

$$\int_{-1}^{1} v_j^{(0)} \nabla^2 \psi_l^{(0)}\, dy = a_j \int_{-1}^{1} v_j^{(0)} \psi_l^{(0)}\, dy \tag{8.60}$$

may also be proved. Here and below, the subscripts take on integral positive values that correspond to the eigenvalues and a_j, b_j numbered in the order their value increases; δ_{jl} is the Kroneker symbol.

The range of solutions to each equation in (8.58), satisfying boundary conditions (8.57), falls into two groups, namely, even and odd. For even perturbations of axial velocity and the stream function we have

$$v_n^{(0)} = \cos \frac{\pi}{2}(n+1)y\ , \quad \psi_n^{(0)} = \frac{1}{\sqrt{I_n}}\left(\frac{\operatorname{ch} \kappa y}{\operatorname{ch} \kappa} - \frac{\cos \delta_n y}{\cos \delta_n}\right)^{*} \tag{8.61}$$

where

$I_n = (1 + \kappa^2/\delta_n^2)(\delta_n^2 - \kappa \operatorname{th}\kappa - \kappa^2 \operatorname{th}^2\kappa)$; δ_n are the radicals of the equation

$$\delta_n \operatorname{tg} \delta_n = -\kappa \operatorname{th}\kappa,\ n = 0, 2, 4\ \ldots$$

numbered in ascending order. The odd eigenfunctions are

$$v_n^{(0)} = \sin \frac{\pi}{2}(n+1)y\ ; \quad \psi_n^{(0)} = \frac{1}{\sqrt{I_n}}\left(\frac{\operatorname{sh} \kappa y}{\operatorname{sh} \kappa} - \frac{\sin \delta_n y}{\sin \delta_n}\right) \tag{8.62}$$

where $I_n = (1 + \kappa^2/\delta n^2)(\delta_n^2 - \kappa \operatorname{cth}\kappa - \kappa^2 \operatorname{cth}^2\kappa)$, δ_n are the radicals of the equation

$$\delta_n \operatorname{ctg} \delta_n = \kappa \operatorname{cth} \kappa$$

* See Nomenclature p. 211 for explanation of notation for trigonometric functions.

numbered in ascending order. The eigenvalues for even and odd perturbations are specified by the relations

$$a_n = \kappa^2 + \frac{\pi^2}{4}(n+1)^2 ; \quad b_n = \kappa^2 + \delta_n^2 , \quad n = 0, 1, 2 \ldots \tag{8.63}$$

We now present the solution to problem (8.56) as the series

$$v = \sum_n A_n v_n^{(0)} ; \quad \psi = \sum_n B_m \psi_m \tag{8.64}$$

After substituting these relations into (8.55) and scalar multiplication of the first equation by $v^{(0)}$ (j assumes the same values as n) and the second one by $\psi_l^{(0)}$ (j assumes the same values as m) we arrive at a system of linear homogeneous equations used to evaluate the coefficients A_n and B_m:

$$\sum_n [a_n(1+Se_z^2) + Se_x^2 \kappa^2 - \sigma]\delta_{in} A_n - \sum_m (Se_z e_x b_j I_{jm} - i\kappa J_{jm}) B_m = 0 ;$$

$$\sum_n Se_z e_x a_n I_{en} A_n + \sum_m [b_m(1+Se_x^2) - \sigma]\delta_{em} B_m = 0 \tag{8.65}$$

Here,

$$I_{jm} = \int_{-1}^{1} v_j^{(0)} \psi_m^{(0)} \, dy , \quad J_{jm} = \int_{-1}^{1} v_j^{(0)} y \psi_m \, dy$$

The set of equations (8.65) represents two independent subsets. One describes perturbations with an even profile v and odd profile ψ; the other describes perturbations with an odd profile v and even profile ψ. The set has a nontrivial solution when its determinant is zero. This condition is prescribed by an algebraic equation whose order relative to σ is equal to the total number of terms in series expansions of (8.64) and, hence, evaluating the same number of coefficients. The range of coefficients may be calculated effectively as a range of eigenvalues of the coefficient matrix in set (8.65). Such a calculation was performed using the first six functions $v^{(0)}$ and six functions $\psi^{(0)}$ for perturbations of each group. With twelve terms in the series expansions of (8.64) preserved, the control calculations gave uncertainties not exceeding 3 per cent for the minimum eigenvalue, which testifies to high efficiency of basis (8.61), (8.62) for an approximate solution of boundary-value problem (8.55) through (8.57).

All the frequencies so calculated are complexes. At small Reynolds numbers R their real part Re $\{\sigma\} > 0$. So, at small R the vibrational perturbations are damping. However, with increasing R a monotone decrease of the real part of the lower level σ is observed both for even and odd perturbations resulting in sign reversal at some R $=$ R(κ,S,α). Here, the imaginary part of the frequency Im $\{\sigma\} = \omega$ is other than zero, which indicates the oscillatory behavior of instability. In Figs. 8.5 and 8.6 neutral curves R(κ) and $\omega(\kappa)$ of odd perturbations at $S = 1$ are presented for different field orientations specified by the angle α between the field and axis z. The curves are seen to be similar to the neutral curves of thermoconvective instability of a fluid layer heated from

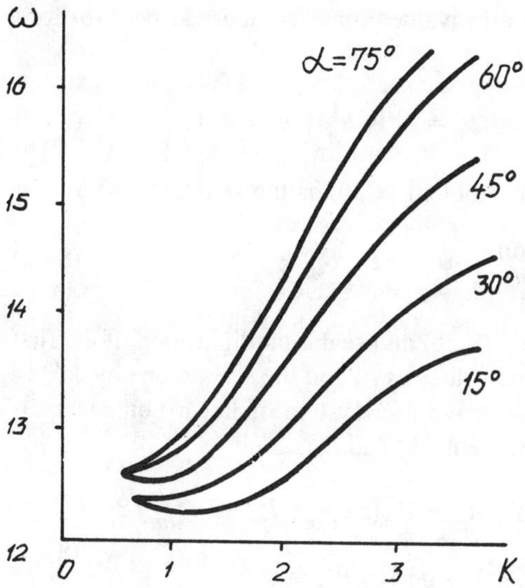

FIG. 8.5

below. The critical Reynolds number $R_* = \min R$ essentially depends on the field orientation. It is minimal at $\alpha = 45°$ and increases as the field direction approaches both the longitudinal and normal-to-the-transverse flow orientations. The critical wave number κ_* depends little on the field orientation. The spatial

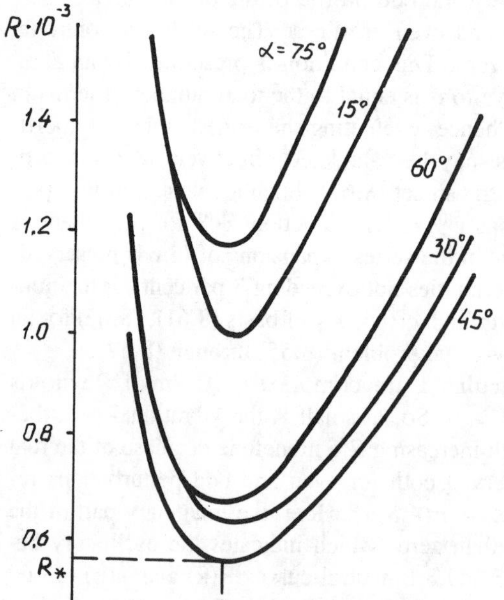

FIG. 8.6

period of a critical perturbation for different α is approximately twice as large as the layer width. The qualitative behavior of the neutral curves does not change with the varying parameter S. The dependence of the critical Reynolds number on S may be seen from Fig. 8.7 plotting R(S) for different α and $\kappa = 1.5 \approx \kappa$. With the magnetic field off, when $S \to 0$, the critical number $R_* \to \infty$. This conclusion agrees with the known result concerning absolute stability of viscous flows with respect to perturbations whose characteristics are constant throughout the stream.

The mechanism of the instability found may exert essential influence on the MF flow in magnetic fields as it manifests itself much earlier before the normal mechanism of hydrodynamic instability. As is known, the linear theory of viscous fluid stability to perturbations whose characteristics do not change in the direction normal to the flow, gives for plane-parallel flow the critical $R_* = 11600$. The measured critical Reynolds number is approximately six times as small as the one predicted by the theory. Following the calculations, presented in Fig. 8.7, R_* for the instability observed at $\alpha = 45°$ and $S \geqslant 0.2$ is below the experimental level of the linear flow of a normal fluid. It is possible that the considered mechanism will lead to a greater decrease of R_* for perturbations of a more complex structure.

The neutral curve for the most detrimental even perturbation is shown in Fig. 8.8. It is seen to be similar to the neutral curve for odd perturbations, the critical R_* being much higher and, hence, the straight-line flow is destroyed by odd perturbations as R increases.

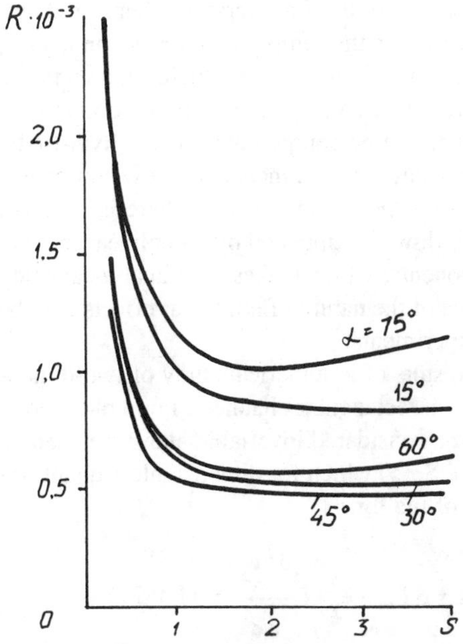

FIG. 8.7

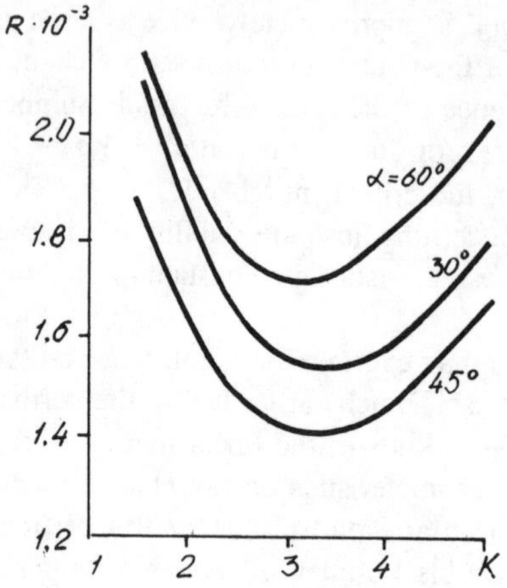

FIG. 8.8

The numerical experimental results in the next section support the results of the linear analysis cited above.

8.4 SWIRLED RECTANGULAR CHANNEL FLOW DUE TO PRESSURE DROP

The physical meaning of no-excitation of a plane flow by an axial one (8.12) implies that the volume force normal to the channel axis and caused by transverse stresses can be completely balanced by the transverse pressure gradient. Such equilibrium may only be achieved for the simplest geometry and, as is shown in the previous section, in a limited range of Reynolds numbers. In channels of a more complex geometry, the transverse force field has both potential and solenoid components. The latter can be compensated by the hydrostatic pressure at none of the R values. As a result, a two-dimensional flow component appears at whatever small pressure drops there may be, i.e., there is no swirl flow threshold. The specificity of such flows is stipulated by a nonlinear interaction of the plane and axial flow components, which makes their analytical study a difficult undertaking. An integral idea of the nature of nonlinear effects may be obtained on the basis of numerical experiments.

The present section presents the results of a numerical study of two-dimensional flows caused by a pressure drop in rectangular channels. For a rather long channel, the flow characteristics may be considered invariable along the channel axis. They may be described by model (8.15) which for dimensionless quantities in the Cartesian coordinate system is of the form:

$$\frac{\partial v}{\partial t}+\frac{\partial \psi}{\partial y}\frac{\partial v}{\partial x}-\frac{\partial \psi}{\partial x}\frac{\partial v}{\partial y}=-E+[1+S(1-e_y^2)]\frac{\partial^2 v_x}{\partial x^2}+[1+S(1-$$

$$-e_x^2)]\frac{\partial^2 v}{\partial y^2}+2Se_xe_y\frac{\partial^2 v}{\partial x\partial y}-S(e_x\frac{\partial}{\partial y}\nabla^2\psi-e_y\frac{\partial}{\partial x}\nabla^2\psi); \tag{8.66}$$

$$\frac{\partial}{\partial t}\nabla^2\psi+\frac{\partial\psi}{\partial y}\frac{\partial}{\partial x}\nabla^2\psi-\frac{\partial\psi}{\partial x}\frac{\partial}{\partial y}\nabla^2\psi=[1+S(1-e_z^2)]\nabla^4\psi+$$

$$+Se_z(e_x\frac{\partial}{\partial y}\nabla^2 v-e_y\frac{\partial}{\partial x}\nabla^2 v)$$

Here $E = \rho p_* y_*^2/\eta^2$; $S = \eta_r/\eta$. The measurement units are: channel width y_* for distance; times y_*^2/ν, velocities ν/y_*; pressure drop, p_*, between two points along the channel axis y_* part for pressure.

Rectangular Channel Flow

In this case, the boundary conditions for fluid adhesion and no-flow are:

$$\begin{aligned} v = \psi = \partial\psi/\partial y = 0 \quad &\text{at } y = 0;\ 1 \\ v = \psi = \partial\psi/\partial x = 0 \quad &\text{at } x = 0;\ 1 \end{aligned} \tag{8.67}$$

The problem geometry and the trajectories of particles in a swirl motion, are shown in Fig. 8.9. The flow mode is determined by four independent parameters, E, S and angles α and β specifying the flow orientation. The set of equations (8.66) is almost similar to the set of equations of convective heat transfer for plane flows and may, therefore, be solved in the manner the convective heat transfer problems were solved. Problem (8.66), (8.67) was solved by the net-point method using a longitudinal-transverse monotone conservative difference network of the second order of accuracy on a spatial grid of 1/20 spacing. In the considered range of parameters the flow was steady. The structure of the isolines

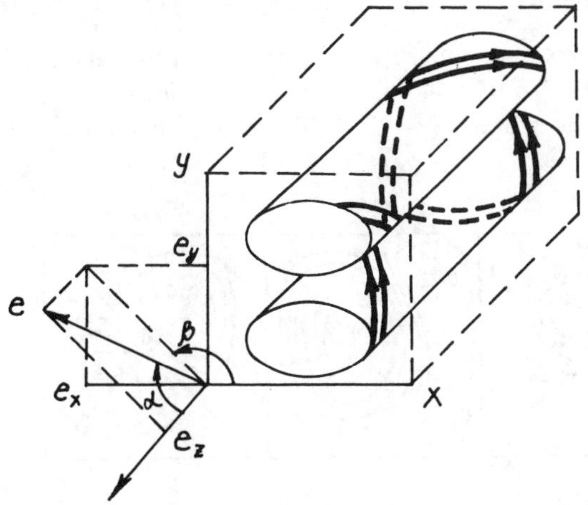

FIG. 8.9

of axial velocity and stream function was studied as well as the main flow characteristics, such as maximum axial velocity v_m, stream function ψ_m and fluid flow rate Q depending on the governing parameters.

If the field direction and the channel axis make an angle different from 0 and $\pi/2$, then the stream function is not zero at any values of the parameters E and S. At small pressure drops ($E \leqslant 10^4$), however, the plane flow component is of low intensity and slightly affects the axial flow characteristics. The stream function isolines for different angles $0 < \beta < 45°$ at fixed values of parameters $\alpha = \pi/4$; $S = 1$; $E = 10^4$, are presented in Fig. 8.10. Taking account of the problem symmetry in this figure, we may get a notion about a plane flow component at any values of the angle β. The impact of the other parameters on flow characteristics were studied for $\beta = 0$, since this is the case when the structure of the plane components provides the most intensive "stall" of the axial flow and, consequently, the effects of nonlinear interaction may be most distinct.

In Fig. 8.11, v_m (curve 1), Q (curve 2) and ψ_m (curve 3) are plotted depending on the dimensionless pressure drop E at fixed $S = 1$, $\alpha = \pi/4$. A sharp increase of ψ_m within $E = 10^4 \div 5 \cdot 10^4$ points to the crisis of this dependence. The enhancement of the plane flow component leads to an essential rearrangement of the axial flow. In the supercritical region $E \geqslant 5 \cdot 10^4$, its profile has two maxima which, as E increases, become more and more distinct and may form a "two-hump" profile. This can be easily seen in Fig. 8.12 depicting the isolines for v and ψ at $\alpha = \pi/4$, $S = 1$ for different values of E. The crisis phenomena may be observed experimentally from the flow vs pressure drop bending curve. It is linear in the subcritical region, then the power law with an exponent less than unity is effective in the supercritical region. In this case, the exponent for v_m equal to 0.79 is somewhat less than that for Q equal to 0.83.

The dependences of flow characteristics on the angle for $S = 1$ and two values of the parameter E are presented in Fig. 8.13. Curves 1, 2 and 3 are for v_m, Q and ψ_m at $E = 10^4$; curves 1′, 2′ and 3′ are for $\frac{1}{5}v_m$, $\frac{1}{5}Q$ and $\frac{1}{5}\psi_m$ at $E = 5 \cdot 10^4$. It is seen that maximum intensity of plane motion for subcritical pressure drops is observed at $\alpha = \pi/3$; and v_m, Q are seen to increase monotonely with α growing from 0 to $\pi/2$. At $E = 5 \times 10^4$, the maximum intensity

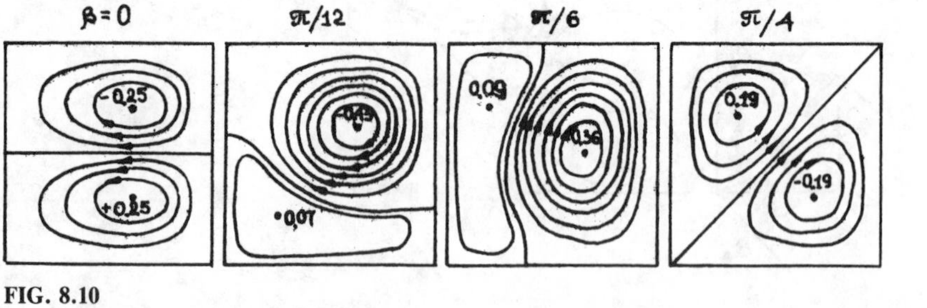

FIG. 8.10

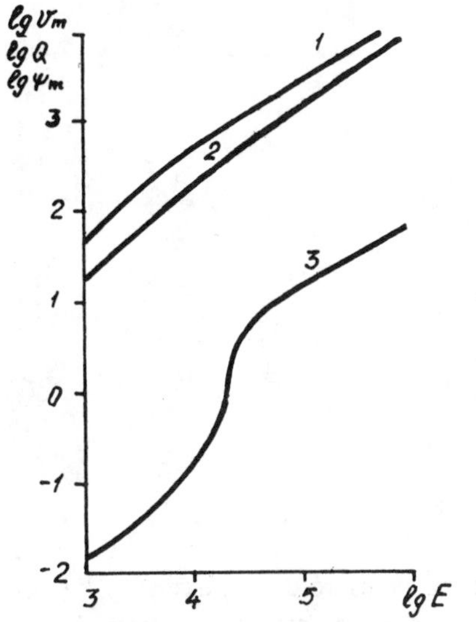

FIG. 8.11

of plane flow is at $\alpha = \pi/4$; the flow rate curves and v_m are no longer monotone. At $\alpha = \pi/6$, they show a maximum.

The effect of the parameter S on the flow characteristics at $\alpha = \pi/4$, $E = 10^4$, 5×10^4 are illustrated in Fig. 8.14. The notations here are the same as in the previous figure. It is seen that, with growing S, the axial velocity and the flow rate decrease monotonely; the dependence $\psi_m(S)$ has a maximum. The divergence of primed and unprimed curves in Figs. 8.13 and 8.14 characterizes nonlinear effects. They are seen to be most distinct for $S \sim 1$, $\alpha \sim \pi/4$.

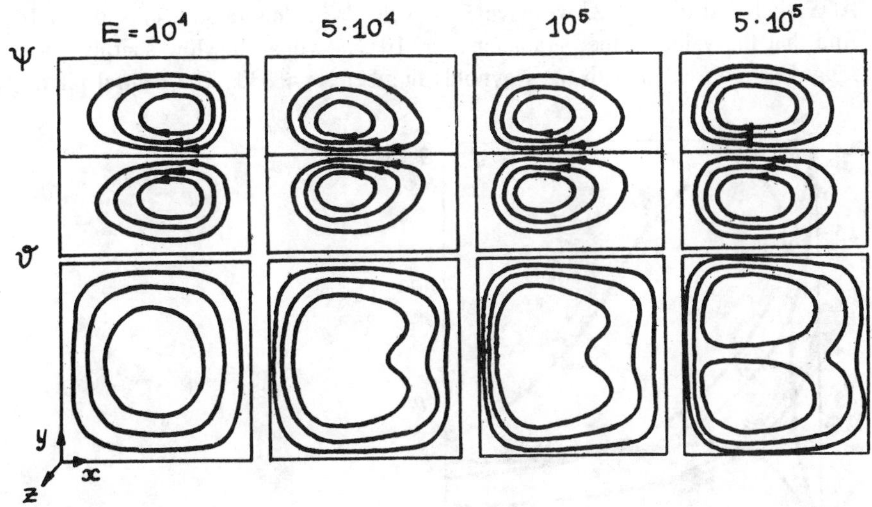

FIG. 8.12

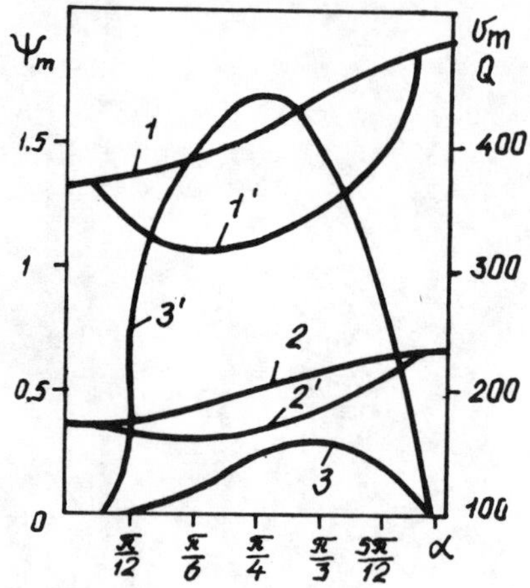

FIG. 8.13

The critical behavior of the flow rate vs. pressure drop curve may be explained by the instability mechanism discussed in the previous section. To do this, we shall relate the parameters E and R. The length scale of sets (8.66) and (8.55) are related as $y_* = 2l$. The characteristic unperturbed flow velocity, entering the Reynolds number, is related with the pressure scale of set (8.66) as $v_0^* \dfrac{e^2}{2\eta_e} \cdot \dfrac{p^*}{2l}$. Substitute the expression for y_* and p_* into the relation for E (see (8.66) and so on) and take account of (8.56) to find

$$E = 8[1 + S(1 - e_x^2)]\mathrm{R} \tag{8.68}$$

At $S = 1$ and $e_x^2 = 1/2$, we have $E = 12\mathrm{R}$. With this relationship in mind, find that the value of the parameter $E = 10^4$, at which the flow sharply rearranges, is consistent with the Reynolds number $\mathrm{R} = 830$. The neutral plane-

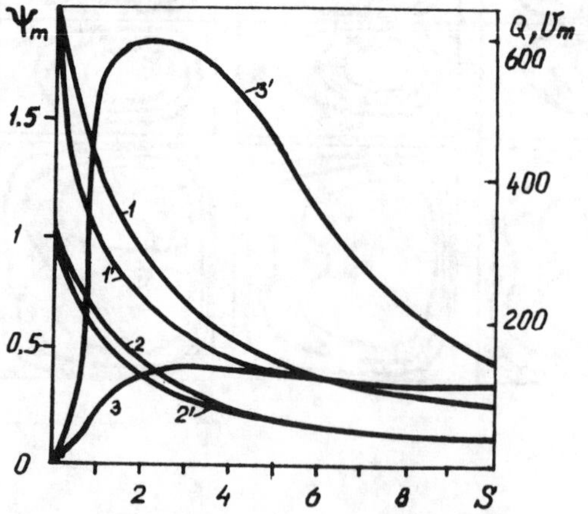

FIG. 8.14

parallel flow instability curve (see Fig. 8.5) at the same S and e_x for a perturbation with the wavelength equal to the layer width, which is most detrimental in case of a rectangular channel flow, is approximately of the same value. The steady rectangular channel flow is likely to be caused by a standing wave formed when the running wave, generated by the flow, is repelled from the walls.

Flow in a Plane-Parallel Layer

The boundary conditions in the planes bounding the layer are included in the first of relations (8.67). We consider the layer to be unbounded in the X-direction. In order to integrate set (8.66) in the finite domain, it is necessary to prescribe the periodic boundary condition. We shall try a solution whose space period in the X-direction is equal to the critical wavelength of the screw instability of the plane-parallel Poiseuille flow which is approximately the double width of the layer. So,

$$\Psi_{x=0} = \Psi|_{x=2};\ v|_{x=0} = v|_{x=2} \tag{8.69}$$

These boundary conditions allowed numerical simulation of the plane-parallel flow, with a rectangular axial velocity profile and zero stream function. Calculations were performed at fixed values of the parameter $S = 1$ in the field oriented in the layer plane ($\beta = 0$) at and angle $\alpha = \pi/4$ to the main stream. At $E < 0.7 \times 10^4$, artificially introduced small perturbations were damped. With the dimensionless pressure drop E exceeding this value, the straight-line flow was seen to collapse. The development of the stream function and axial velocity perturbations in time at $E = 2.75 \times 10^4$ is presented, correspondingly, in Figs. 8.15 and 8.16. Curves 3 are for the dynamics of the process at a point on the channel axis; curves 2, at a point 1/4 channel width distant from the wall. Curves 3 plot maximum volume values of ψ and v vs. time. It can be seen that the development of instability results in a vibrational flow mode. The instantaneous patterns of stream function and axial velocity isolines are depicted in Figs. 8.17 and 8.18.

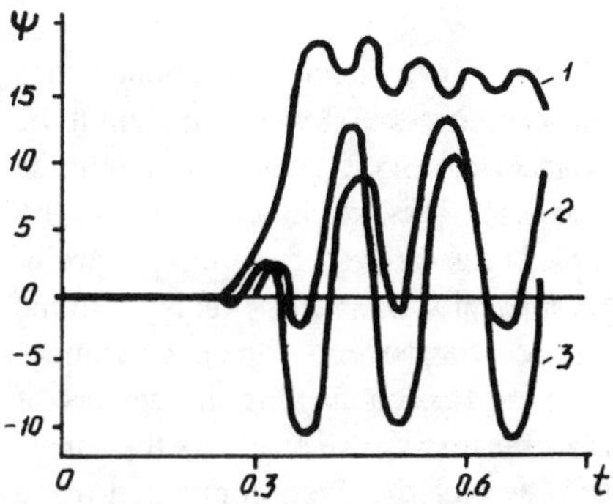

FIG. 8.15

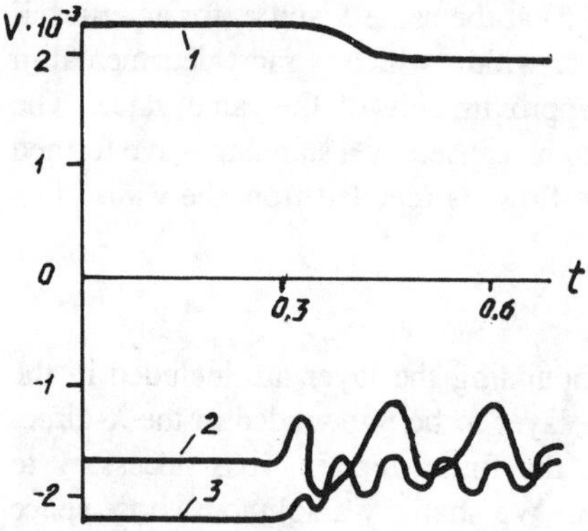

FIG. 8.16

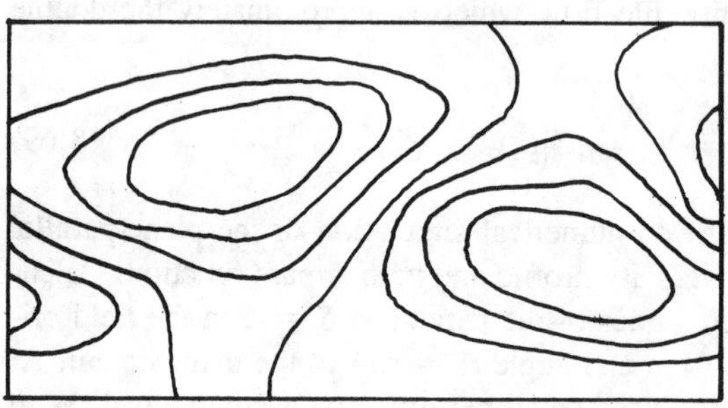

FIG. 8.17

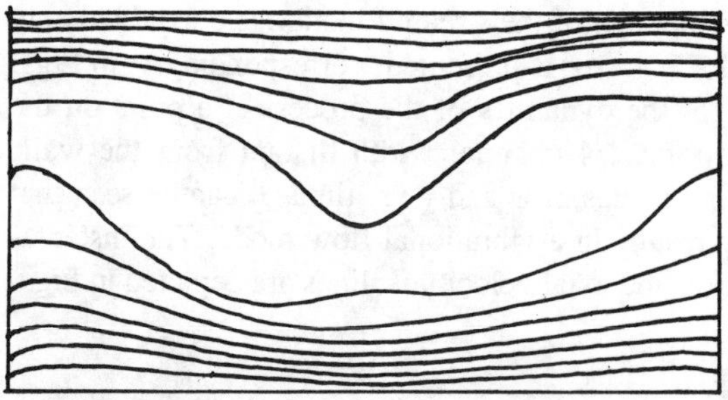

FIG. 8.18

They have been obtained for $E = 2.75 \times 10^4$, which corresponds to a 400% supercritical level. Such a level promotes the plane flow component to be rather developed. This component assures an axial velocity vibration amplitude at a point 1/4 channel width distant from the wall. This amplitude is about 20% of unperturbed axial velocity (cf. Fig. 8.16). It can be seen from the picture of stream function isolines (cf. Fig. 8.17) that ψ distribution across the layer in the far supercritical region has both an even and odd components. The even component has however a large amplitude because the line connecting the centers of two neighbouring cells is closer to the horizontal than the vertical. As the supercritical level decreases, the relative contribution of the even component to ψ

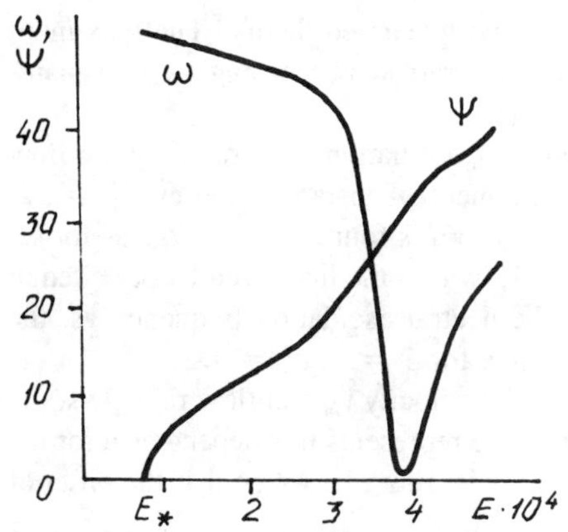

FIG. 8.19

distribution increases. At the initial instant of time of the development of a perturbation, the stream function is an even coordinate function in the vicinity of the critical point. The ψ isoline patterns represent two correct cells, divided by the vertical line, moving from left to right. Thus, the straightline flow is destroyed by even ψ perturbations, which is in agreement with the results of the linearized theory. In the steady mode, a relatively small odd component appears in ψ distribution even at a small supercritical level.

The dependence of the maximum stream function ψ_m on the parameter E is presented in Fig. 8.19 (curve 1). The curve was built by reducing the parameter E. As an initial approximation to the problem solution at some value of E, we used the axial velocity and stream function distribution for a previous and much higher value of E. The critical $E = 0.7 \times 10^4$, at which the plane flow component appears, was obtained both in unperturbed and developed swirl flow approximations. No essential difference in the E values thus obtained was observed. The relationship between the dimensionless pressure drop E and Reynolds number (8.68) makes it possible to ascertain that in the case under consideration the critical Reynolds number is $R_* = 580$. This value agrees well

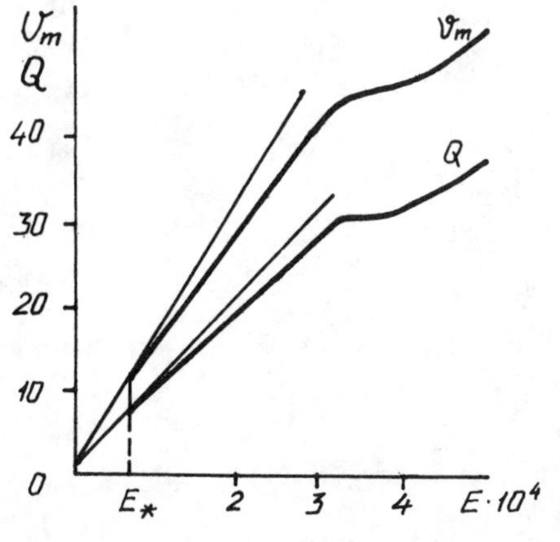

FIG. 8.20

with the results obtained with the use of the linearized theory. The maximum stream function grows with the subcritical parameter according to the square-root law. For the studied version $\psi_m = 5(E - E_*)$.

Curve 2 in Fig. 8.19 is the frequency of oscillations generated by the flow vs. parameter E. It is seen that at the moment of instability onset $\omega_* = 52$. Having in mind that the time scale for set (8.66) is four times as large as for set (8.55), we find that it consists with $\omega = 13$ within the linearized theory cited in the previous section. In Fig. 8.6 with the neutral oscillation frequency vs. the Reynolds number, we obtain the same value for $S = 1$, $e_x^2 = 1/2$.

Figure 8.20 presents the maximum axial velocity v_m and flow rate Q vs. the dimensionless pressure drop. The dashed line represents this dependence for the plane-parallel flow; the solid line, for the swirl flow. It is seen that the onset of the plane flow component decreases v_m and Q.

CHAPTER

NINE

DYNAMIC INTERACTION BETWEEN MAGNETIC FLUIDS AND A NONUNIFORM FIELD

9.1 EQUILIBRIUM OF A MAGNETIC FLUID ACTED UPON A DYNAMIC FORCE

When some vessel with MF moves in a nonuniform nonstationary field, the fluid is under the action of volume force (6.36) due to the dynamic magnetization component which may cause the fluid to mix. We shall now elucidate the possibility of fluid equilibrium under the action of this force. The equilibrium condition implies potentiality of the volume force: $\nabla \times -\vec{F} = 0$ at $\vec{v} = 0$. Transform (6.36) for a nonconducting fluid ($\vec{j} = 0$) taking account of the equations of field, continuity and equation for $\vec{M}'$. We shall only consider the interaction of the fluid and external field which are described by the Maxwell equations in the noninductive, approximation: $\nabla \times \vec{H} = 0$, $\nabla \cdot \vec{H} = 0$. For such fields, the second term in (6.36) can be presented in the form

$$(\vec{M}' \cdot \nabla)\vec{H} = 0.5[\nabla \times (\vec{M}' \times \vec{H}) - \vec{H}\nabla \cdot \vec{M}' - \vec{H} \times \nabla \times \vec{M}' - \nabla(\vec{M}' \cdot \vec{H})]$$

and for dynamic volume force we can obtain the expression

$$\vec{F}' = 0.5\mu_0[\nabla(\vec{M}'\vec{H}) - \vec{H}\nabla \cdot \vec{M}' - \vec{H} \times \nabla \times M'] \qquad (9.1)$$

If dynamic magnetization is represented as a product of some scalar a and vector $\vec{D}$ parameters, i.e. $\vec{M}' = \alpha \vec{D}$, then (9.1) takes on the form

$$\vec{F}' = 0.5\mu_0\{\alpha[\nabla(\vec{H}\cdot\vec{D}) - \vec{H}\nabla\vec{D} - \vec{H}\times(\nabla\times\vec{D})] - \nabla\alpha\times(\vec{H}\times\vec{D})\} \tag{9.2}$$

For the component $\vec{M}'$ parallel to the field, we shall set here $\vec{D} \equiv \vec{H}$, and in this case $\vec{F} = \mu \cdot aH\nabla H$. Using this relation along with (9.2) and the equation for $\vec{M}'$, we find

$$\mu_0^{-1}\vec{F}' = -\Delta\kappa\dot{H}\nabla H - 0{,}5\nabla\kappa_\perp \times[\vec{H}\times\dot{\vec{H}} - H^2\vec{\Omega} + H(\vec{\Omega}\cdot\vec{H})] -$$
$$- 0.5\kappa_\perp[\nabla(H\cdot\dot{H}) - \vec{H}\nabla\cdot(\dot{\vec{H}} - \vec{\Omega}\times\vec{H}) - \vec{H}\times\nabla\times(\dot{\vec{H}} - \vec{\Omega}\times\vec{H})] \tag{9.3}$$

Here, it should be taken into account that the total derivative $\vec{H}$ under the operation sign ∇ and $\nabla\times$, are equal to the convective part as $\nabla\cdot\partial\vec{H}/\partial t = \nabla\times\partial H/\partial t = 0$.

We shall now derive an expression for dynamic force in an incompressible fluid. Considering the continuity equation $\nabla\cdot\vec{v} = 0$, we may arrive at the relation

$$(\vec{v}\cdot\nabla)\vec{H} = 0.5(\nabla\times(\vec{H}\times\vec{v}) + \nabla(\vec{H}\cdot\vec{v})$$

Hence it follows that

$$\nabla\cdot(\dot{\vec{H}} - \vec{\Omega}\times\vec{H}) = 0{.}5\nabla^2(\vec{H}\cdot\vec{v});$$
$$\nabla\times(\dot{\vec{H}} - \vec{\Omega}\times\vec{H}) = -0{.}5\nabla^2(\vec{H}\times\vec{v}) - \nabla(\vec{H}\cdot\vec{\Omega})$$

Consequently,

$$\mu_0^{-1}\vec{F}' = -\Delta\kappa\dot{H}\nabla H - 0.5(\nabla\kappa_\perp\times(\vec{H}\times\dot{\vec{H}} - H^2\vec{\Omega}) - 0{,}5\ \vec{H}\times\nabla(\kappa\vec{H}\vec{\Omega}) -$$
$$- 0.5\kappa_\perp[\nabla\ (\dot{H}H) - 0{,}5\vec{H}\nabla^2(\vec{H}\cdot\vec{v}) + 0{,}5\vec{H}\times\nabla^2(\vec{H}\times\vec{v})] \tag{9.4}$$

For a steady fluid exposed to a nonstationary field we have from (9.3), (9.4)

$$\mu_0^{-1}\vec{F} = -\Delta\kappa\frac{\partial H}{\partial t}\nabla H - \frac{1}{2}\kappa_\perp\nabla(\frac{\partial H}{\partial t}H) - \nabla\kappa_\perp\times(\vec{H}\times\frac{\partial\vec{H}}{\partial t}) \tag{9.5}$$

Hence it follows that the dynamic susceptibility tensor is isotropic ($\Delta\kappa = 0$) and uniform ($\kappa_\perp =$ const); the dynamic force is potential

$$\vec{F}' = 0{,}5\mu_0\kappa_\perp\nabla(H\frac{\partial H}{\partial t}) \tag{9.6}$$

These conditions are satisfied at $H \ll H_T$, when $\kappa_\parallel = \kappa_\perp = \kappa_0 =$ const. Thus, in "weak" magnetic fields of an arbitrary configuration the dynamic force is balanced by the pressure gradient, and the fluid is in equilibrium. If $H \gtrsim H_T$, the isotropy and uniformity of the dynamic susceptibility tensor is disturbed with the result that the nonstationary field is responsible for convective mixing of the fluid. Below is the expression for F' in "strong" ($H \gg H_T$) fields when $\kappa_\parallel = 0 \mu_0 \kappa_\perp H^2/4 = \eta_{rs} =$ const:

$$\vec{F}' = 2H^{-1}\eta_{rs}\left[\frac{\partial H}{\partial t}\nabla H - H\nabla\frac{\partial H}{\partial t} - 2H^{-1}\nabla H \times \left(\vec{H} \times \frac{\partial \vec{H}}{\partial t}\right)\right] \tag{9.7}$$

The study of the effect of this force on potentiality cannot yield general results. However, for some partial geometries (see, for example, Sec. 9.2) a magnetic field may also be in equilibrium in "strong" fields. It should be noted that macroscopic motion of a magnetic fluid may be caused by its concentration nonuniformity and nonisothermality, since the dynamic susceptibility coefficients depend on the ferrophase concentration and temperature and on the field intensity into the bargain.

The conclusion about potentiality of the dynamic force in weak fields of an arbitrary configuration may be generalized for the case when the fluid moves as a solid. The most general expression for velocity distribution in the case of a quasisolid flow is of the form $\vec{v} = \vec{v}^* + \vec{\Omega}^* \times \vec{r}$. Here $\vec{v}^*$, $\vec{\Omega}^*$ are the velocities of translational and rotational fluid motions which are constant for the entire fluid. Direct calculation in the Cartesian coordinate system may show that, in this case, the two last terms in (9.3) are zero and, therefore,

$$\mu_0^{-1}\vec{F}' = -\Delta\kappa\dot{H}\nabla H - 0.5\kappa_\perp \nabla(HH) \div 0.5\nabla\kappa_\perp \times [\vec{H}\times\dot{\vec{H}} - H^2\Omega^* + \\ + \vec{H}(\vec{\Omega}^*\cdot\vec{H})] \tag{9.8}$$

For "weak" fields it gives the relation

$$\vec{F}' = -0.5\,\mu_0\kappa_0\nabla\{\partial H/\partial t + [(\vec{v}^* + \vec{\Omega}^* \times \vec{r})\nabla]H\} \tag{9.9}$$

from which (9.6) follows as a partial case. Thus, dynamic interaction does not disturb the state of rest or the state of motion of a homogeneous isotropic MF as a solid in an external field of arbitrary configuration. This conclusion also holds in taking account of the fields induced by equilibrium magnetization provided by magnetic susceptibility of the fluid is constant, since in this case, too, there are no volume sources in the field equations. The MF motion under the action of nonstationary nonuniform fields, observed experimentally, should be attributed to the perturbation of MF homogeneity and isotropy.

9.2 INTERACTION BETWEEN MAGNETIC FLUID AND ONE-SIDED INDUCTOR FIELD

We shall now discuss in detail the dynamic interaction of MFs with a nonuniform field for plane-parallel flows in the field of a one-sided magnetic inductor.

The idealized magnetic inductor is a unidirectional surface current whose density varies in the direction normal to the current. Let us have the axis z of the Cartesian coordinate system be in the current direction; and the axis y normal to the current surface. The components and the modulus of intensity of the running field induced in the half-space $y \geqslant 0$ by current $I_z = I_* \cos(\delta x - \omega t)$ harmonically changing in time and along the coordinate x are of the form

$$H_x = -H_* e^{-\delta y} \cos(\delta x - \omega t); \; H_y = H_* e^{-\delta y} \sin(\delta x - \omega t)$$
$$H = H_* e^{-\delta y} \tag{9.10}$$

Here, $H_* = I_* \dfrac{\mu^2}{\mu_1 + \mu_2}$; μ_2, μ_1 are magnetic susceptibilities of the substances filling the spaces $y < 0$ and $y > 0$, respectively. Assume that the magnetic fluid fills a plane-parallel layer of thickness d. The problem geometry is presented in Fig. 9.1.

In a noninductive approximation, the field in the fluid is specified by relation (9.10) where $\mu_1 = 1$. We may show that such a field obeys the relation

$$(\vec{M}\nabla)\vec{H} = -\delta(\vec{M} \cdot \vec{H})\vec{i}_y + \delta(\vec{M} \times \vec{H})\vec{i}_x \tag{9.11}$$

It is seen from (9.10) that the field intensity is only dependent on the coordinate y. In this connection, we shall consider plane-parallel flows for which the velocity distribution depends only on the coordinate y: $v_x = v(y, t)$. In this case, dynamic force (6.36) may be presented, with regard to (9.11), as

$$\vec{F}' = (\partial \delta'^{a}_{xy}/\partial y + 2\delta\sigma'^{a}_{xy})\vec{i}_x \tag{9.12}$$

where $\sigma^a_{xy} = 0.5\,\mu_0(\vec{M} \times \vec{H})_z$. The equation of motion (6.38) in the projections onto the coordinate axes becomes:

$$\rho \frac{\partial v}{dt} = -\frac{\partial p}{\partial x} + \frac{\partial}{\partial y}(\sigma'^{s}_{xy} + \sigma'^{a}_{xy}) + 2\delta\sigma'^{a}_{xy} \tag{9.13}$$

$$\partial p/\partial y = -\mu_0 \delta M_0 H \tag{9.14}$$

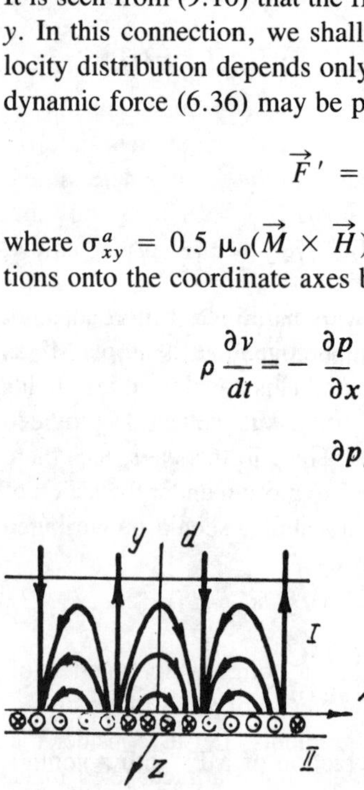

FIG. 9.1

where

$$\sigma'^{s}_{xy} = \eta \partial v/\partial y , \quad \sigma'^{a}_{xy} = \eta_r \, (2(\vec{e} \times \dot{\vec{e}})_z + \partial v/\partial y)$$

As $e_x = -\cos(\delta x - \omega t)$, $e_y = \sin(\delta x - \omega t)$, for the field rotational velocity relative to the material element in translational motion we have the expression $(\vec{e} \times \dot{\vec{e}})_z = -\delta(v - v_{\mathrm{m}})$. Here, $v_\Phi = \omega/\delta$ is the phase velocity of the running field. So, for a small material element in a plane-parallel flow, field (9.10) seems to be constant in modulus and to be rotating at an angular velocity proportional to the relative velocity of the element and the field. Here, if the field is spread in the positive X-direction, it rotates counterclockwise relative to the material element at rest. In case of a steady-state motion, the field rotational velocity is constant for a fixed element but different for different elements if their velocities are different.

For evaluating the velocity profile we have from (9.13) the equation

$$\rho \frac{\partial v}{\partial t} = -\frac{\partial p}{\partial x} + \frac{\partial}{\partial y}[(\eta + \eta_r)\frac{\partial v}{\partial y}] + 2\delta \, (\frac{d\eta_r}{dy} + 2\eta_r \, \delta) \, (v_\Phi - v) \qquad (9.15)$$

The viscous stress component, determining the force of the fluid acting upon the layer boundaries,

$$\sigma'_{xy} = (\eta + \eta_r)\partial v/\partial y - 2\eta_r\delta(v - v_\Phi) \qquad (9.16)$$

Equation (9.15) is characterized by a volume force proportional to the relative velocity of the material element $v - v_\Phi$. This means that propagation of the fluid relative to the walls of the channel with fluid may become a source of magnetic fluid flow or head. The qualitative character of the force is determined by the sign of the quantity $f = 2\delta(\partial\eta_r/dy + 2\eta_r\delta)$. If $f > 0$, it inhibits the relative motion of the material element and the field, if $f < 0$, it promotes the motion.

The derivative $d\eta_r/dy$ may be specified by various physical mechanisms, such as gradients of ferrophase concentration, temperature, field intensity, i.e.

$$\frac{d\eta_r}{dy} = \frac{\partial\eta_r}{\partial\varphi}\frac{\partial\varphi}{\partial y} + \frac{\partial\eta_r}{\partial T}\frac{\partial T}{\partial y} + \frac{\partial\eta_r}{\partial H}\frac{\partial H}{\partial y}$$

For an isothermal fluid of uniform concentration we obtain

$$f = 2\delta^2(2\eta_r - H\partial\eta_r/\partial H) = 0.5\,\delta^2 H^3 \partial\kappa_\perp/\partial H \qquad (9.17)$$

Following (6.24), the dynamic susceptibility $\kappa_\perp$ is a nongrowing function of field intensity and, consequently, $f > 0$. The sign reversal of f may occur owing to the ferrophase temperature and concentration gradients. Let us consider the concentration mechanism. The equilibrium magnetic force attracts ferro magnetic particles to the plane $y = 0$ where the field is maximum. This gives rise to an adverse particle concentration gradient $\partial\varphi/\partial y$. As $\partial\eta_r/\partial\varphi$ is a positive value, the concentration nonuniformity makes an adverse contribution to the coefficient f and may invert the direction of the dynamic volume force. Below, the fluid is assumed to be isothermal and uniform in concentration.

Dynamic Interaction With a "Weak" Field

If $H_* \ll H_T$, the condition of "weak" fields is satisfied in the fluid volume. In this case $\kappa_\perp = \text{const}$. Let us consider a shear flow between the plane-parallel boundaries moving at different velocities:

$$v|_{y=0} = v_0, \; v|_{y=d} = v_0 + v_1 \tag{9.18}$$

Upon single integration, equation (9.15) for a steady flow assumes the form

$$(\eta + \eta_r)\,\partial v/\partial y = y\partial p/\partial x + \text{const} \tag{9.19}$$

Here

$$\eta_r = (1/4)\,\kappa_0 H_*^2\, e^{-2\delta y} = \eta_{r*}\, e^{-2\delta y}$$

Distributions of velocity and viscous stresses, satisfying (9.18) and (9.19), in the absence of a pressure gradient are of the form

$$v = v_0 + v_1 \frac{\delta y + \ln\sqrt{1 + S\,e^{-2\delta y}} - \ln\sqrt{1+S}}{\delta d + \ln\sqrt{1 + S e^{-2\delta d}} - \ln\sqrt{1+S}} \tag{9.20}$$

$$\sigma_{xy} = \eta v_1 \delta \frac{1 - 2S\,(\delta y + \ln\sqrt{1 + S e^{-2\delta y}} - \ln\sqrt{1+S})\; e^{-2\delta d}}{\delta d + \ln\sqrt{1 + S e^{-2\delta y}} - \ln\sqrt{1+S}} \tag{9.21}$$

$-2\eta_{r*}\,\delta e^{-2\delta y}(v_0 - v_\Phi)$, where $S = \eta_{r*}/\eta$

If the layer boundaries move at equal velocities ($v_1 = 0$), it follows from (9.20) that shear flows do not arise in a fluid, which agrees with the general conclusion made in Sec. 9.1. Interaction between the fluid and the field is reduced to energy dissipation in microvortices. Besides, the layer boundaries are affected by the forces due to viscous antisymmetric stresses:

$$\sigma'^a_{xy}\,|_{y=0} = 2\eta_{r*}\delta\,(v_\Phi - v_0); \quad -\sigma'^a_{xy}\,|_{y=d} = -2\eta_{r*}\delta\,(v_\Phi - v)\exp(-2\delta d) \tag{9.22}$$

The mechanism of these forces was discussed in detail in 7.2 for a fluid rotating as a solid in a uniform rotating field. The case is distinctive for the fact that rotation of the field relative to the material element is achieved in case of a translational displacement of the material element and field configuration due to tortuosity of the field lines.

As the field intensity and the rotational viscosity decrease exponentially with the distance from the current space, the force acting upon the boundary $y = 0$ is always greater in modulus than the force acting upon the boundary $y = d$. These forces act in opposition. Viscous stresses tend to retard the motion of the boundary $y = 0$ relative to the field and to accelerate the motion of the boundary $y = d$. The force acting per unit of the current surface due to interaction with fluid, is equal to $\sigma'^{a}_{xy}|_{y=d} - \sigma'^{a}_{xy}|_{y=0}$ and always hampers the displacement of field configuration relative to the fluid.

If any of the boundaries are free, viscous stresses will set it in motion relative to the fixed boundary. In the steady mode, the shear rate near the free boundary will be of the magnitude enough to balance symmetric and antisymmetric stresses. Let us consider this effect for the case when the boundary $y = d$ is free. We shall find its relative velocity from the condition $\sigma_{xy}|_{y=d} = 0$

$$\nu_1 = \frac{2(\nu_0 - \nu_\Phi)S[\delta d + \ln\sqrt{1+S\mathrm{e}^{-2\delta d}} - \ln\sqrt{1+S}]\mathrm{e}^{-2\delta d}}{1 - 2S(\delta d + \ln\sqrt{1+S\mathrm{e}^{-2\delta d}} - \ln\sqrt{1+S})\ \mathrm{e}^{-2\delta d}} \tag{9.23}$$

At $S \ll 1$, we arrive at a simple relation

$$\nu_1 = 2S(\nu_0 - \nu_\Phi)\delta d \exp(-2\delta d) \tag{9.24}$$

In accordance with the direction of antisymmetric stress forces, the unfixed boundary moves in the direction opposite to field propagation. At $2\delta d = 1$, the relative velocity achieves the maximum $\nu_{1m} = S(\nu_0 - \nu_\Phi)/\mathrm{e}$.

Interaction Between Magnetic Fluid and "Strong" Fields

With increasing field intensity, the parameter f (9.17) for the volume dynamic force grows. In "strong" fields, when the rotational viscosity coefficient achieves saturation $\eta_r = \eta_{rs}$, and $H\partial\eta_r/\partial H \to 0$, the coefficient f is maximum. At $H_* \mathrm{e}^{-\delta d} \gg H_\mathrm{T}$, the condition of "strong" fields is satisfied for the whole volume occupied by the fluid. In this case, (9.15) gives for a steady flow the equation

$$(\eta + \eta_{rs})\nu'' + 4\delta^2\eta_r(\nu_\Phi - \nu) = \partial p/\partial x \tag{9.25}$$

Distribution of the velocity and viscous stresses, satisfying (9.25) and boundary conditions (9.18), are of the form

$$\frac{\nu - \nu_\Phi}{\nu_0 - \nu_\Phi} = (1+E)\frac{\mathrm{sh}(2\sqrt{S'}\,\delta y) + \mathrm{sh}[2\sqrt{S'}\,\delta(d-y)]}{\mathrm{sh}(2\sqrt{S'}\,\delta d)}^{*} +$$

* See Nomenclature p. 211 for explanation of notation for trigonometric functions.

$$+ \frac{\nu_1}{\nu_0 - \nu_\Phi} \frac{\mathrm{sh}(2\sqrt{S'}\,\delta y)}{\mathrm{sh}(2\sqrt{S'}\,\delta d} - E \tag{9.26}$$

$$\sigma_{xy} = 2\eta_{rs}\delta(\nu_0 - \nu_\Phi)\left(\frac{1+E}{\sqrt{S'}} \frac{\mathrm{ch}(2\sqrt{S'}\,\delta y) - \mathrm{ch}[2\sqrt{S'}\delta(d-y)]}{\mathrm{sh}(2\sqrt{S'}\delta d)} + \right.$$
$$\left. + \frac{\nu_1}{(\nu_0 - \nu_\Phi)\sqrt{S'}} \frac{\mathrm{ch}(2\sqrt{S'}\delta y)}{\mathrm{sh}(2\sqrt{S'}\delta d)} - \frac{\nu - \nu_\Phi}{\nu_0 - \nu_\Phi} \tag{9.27}$$

Here, $S' = \dfrac{\eta_{rs}}{\eta + \eta_{rs}}$, $E = \dfrac{1}{4\delta^2\eta_{rs}(\nu_0 - \nu_\Phi)} \dfrac{\partial p}{\partial x}$

Consider the interaction of the running field and the fluid in a channel whose boundaries move at equal velocities. The condition of "no fluid flow rate" is satisfied in a closed channel whence it follows that $E = -1$ or

$$\partial p/\partial x = 4\delta^2\eta_{rs}(\nu_\Phi - \nu_0) \tag{9.28}$$

This case is the demonstration of a possible magnetic fluid equilibrium with a "strong" magnetic field applied. The dynamic force is balanced by the pressure gradient, and there is no shear flow. Viscous stresses are distributed nonuniformly:

$$\sigma_{xv} = 2\eta_{rs}\delta(\nu_\Phi - \nu_0) \tag{9.29}$$

In an open channel, for which the condition $\partial_p/\partial x = 0$ holds, the dynamic force induces a flow rate. Distribution of the dimensionless velocity $V = v/v_m$, excited by the running field in the channel with fixed boundaries, is presented in Fig. 9.2b for different $S = \eta_{rs}/\eta$ and $N = \delta d$. In narrow channels, for which $\delta d \ll 1$, the field induces a Poiseuille flow $v = 2v_\Phi\delta^2 S' y(d - y)$.

The viscous stress forces acting upon the channel walls are:

$$\sigma_{xy}|_{y=0} = 2\eta_{rs}\delta(\nu_0 - \nu_\Phi)\left[\frac{1 - \mathrm{ch}(2\sqrt{S'}\delta d)}{\sqrt{S'}\,\mathrm{sh}(2\sqrt{S'}\delta d)} - 1)\right] \tag{9.30}$$

$$-\sigma_{xy}|_{y=d} = 2\eta_{rs}\delta(\nu_o - \nu_\Phi)\left[\frac{1 - \mathrm{ch}(2\sqrt{S}\,\delta d)}{\sqrt{S}\,\mathrm{sh}\,(2\sqrt{S'}\delta d} + 1\right] \tag{9.31}$$

The last cofactor in expression (9.30) is negative at any values of the parameters S' and δd. Consequently, the boundary at rest, $y = 0$, is always affected by the force which causes it to move in the direction of field phase propagation. Of

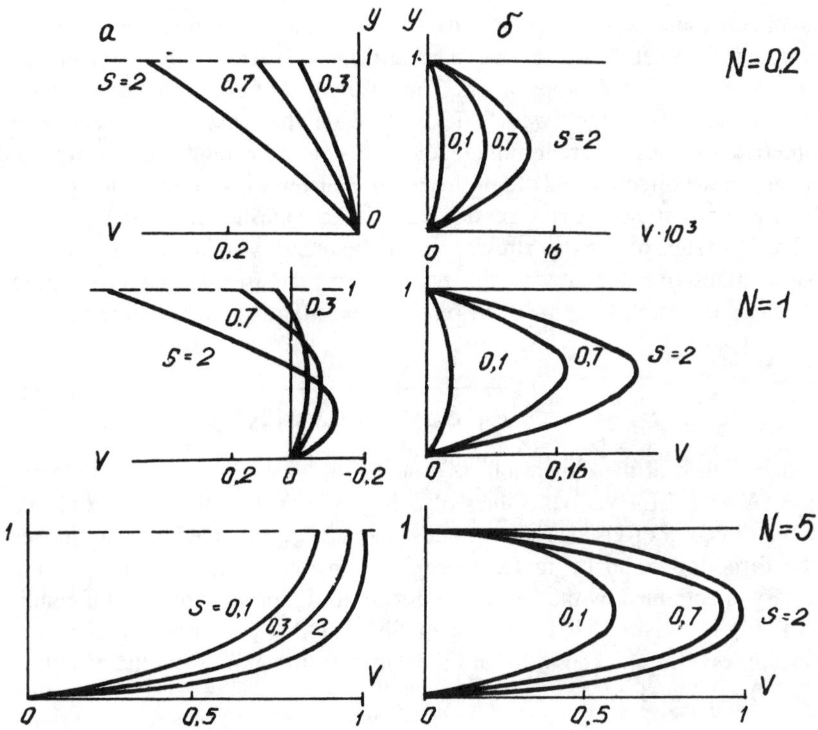

FIG. 9.2

similar nature is the total force $\sigma_{xy}|_{y=0} - \sigma_{xy}|_{y=d}$ equal in sign to the force acting per unit area of the magnetic field source.

The viscous stress force acting upon the boundary $y = d$ at small and large δd is

$$-\sigma_{xy}|_{y=d} = \begin{cases} 2\eta_{rs}\delta(v_0 - v_\Phi)(1 - \delta d/2) & \text{at } 2\delta d \ll 1 \\ 2\eta_{rs}\delta(v_0 - v_\Phi)\dfrac{\sqrt{S'} - 1}{\sqrt{S'}} & \text{at } 2\sqrt{S'}\delta d \gg 1 \end{cases} \tag{9.32}$$

We may conclude from these expressions that, at $\delta d \gg 1$, the running magnetic field strives to entrain the boundary at rest. At $\delta d \ll 1$, the field affects the boundary in an opposite direction. The direction of the force changes provided

$$\mathrm{ch}(2\sqrt{S'}\delta d) - \sqrt{S'}\,\mathrm{sh}(2\sqrt{S'}\delta d) = 1 \tag{9.33}$$

At $S' \ll 1$, this equation has the solution $(\delta d)_* = 1$. With increasing S', (δd) grows as well.

The sign reversal at viscous stresses is attributed to the opposite direction of antisymmetric and symmetric viscous stresses which act upon the boundary $y = d$. The symmetric stresses due to shear flow rate, act upon both boundaries in the direction of field propagation. The particulars of antisymmetric stresses were discussed above for "weak" fields. In wide channels, the mechanism of symmetric stresses is predominant and the net force tends to entrain the boundary in the direction of field propagation. Narrow gaps show predominance of the force of antisymmetric stresses acting in an opposite direction.

That the sign of viscous stresses on the boundary $y = d$ may be reversed imparts qualitative features to the behavior of the free boundary. Setting $\sigma_{xy}|_{y=d} = 0$, from (9.27) we shall find the velocity of the free boundary

$$v_1 = (v_0 - v_\Phi)\left(\frac{1}{\mathrm{ch}(2\sqrt{S'}\delta d) - \sqrt{S'}\mathrm{sh}(2\sqrt{S'}\delta d)} - 1\right) \tag{9.34}$$

The dependence of the dimensionless velocity $V_1 = v_1/v_\Phi(v_0 = 0)$ on the parameter $N = \delta d$ for various values of S is shown in Fig. 9.3. In accordance with the direction of viscous stress forces at small δd, the free boundary moves in the direction opposite to field propagation ($V_1 < 0$). At large δd, the boundary is entrained by the field. The direction of motion is reversed if condition (9.33) is satisfied. It is seen from Fig. 9.3 that, in case of a changing dimensionless film thickness, δd has a negative extremum. From the condition $\partial v_1/\partial(\delta d) = 0$ we find

$$(\delta d)_* = \frac{1}{4\sqrt{S}} \ln \frac{1+\sqrt{S'}}{1-\sqrt{S'}} \tag{9.35}$$

The extreme value of V_1 is

$$V_{1\max} = 1 - \sqrt{1+S} \tag{9.36}$$

Distribution of the dimensionless velocity $V = v'/v_\Phi$ over the film width, plotted according to (9.26) where $E = 0$, $v_0 = 0$ and v_1 is specified by expres-

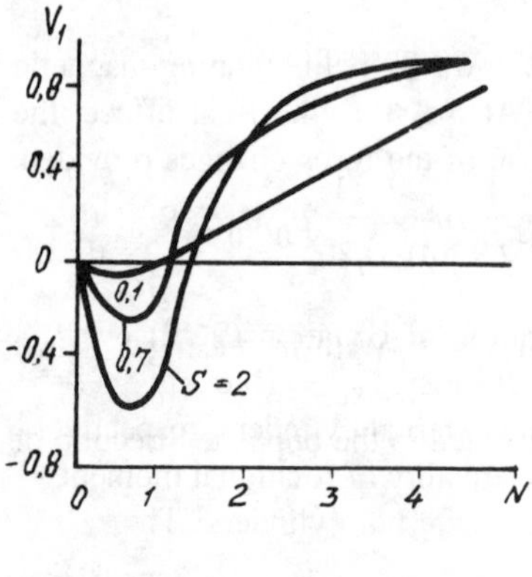

FIG. 9.3

sion (9.34) and presented in Fig. 9.2a. For "thin" films ($\delta d \ll 1$), in a square-small parameter approximation, this distribution is a combination of a linear and square profiles

$$v = -2v_{\Phi}\delta S'[(1 + 2\delta S')y + (y - 2d)\delta y] \tag{9.37}$$

The square profile component is due to the volume force; the linear component provides compensation of film surface symmetric and antisymmetric stresses.

9.3 THE COUETTE FLOW IN RADIAL AND AZIMUTHAL FIELDS

Let us consider the Couette flow of a magnetic fluid in nonuniform fields of cylindrical symmetry which can be conveniently realized in cylindrical rotary viscometers. Such are the field between coaxial radially magnetized cylindrical magnets and the field of a live linear conductor. In the first case, a radial field of a cylindrical monofield may be realized in the gap between the cylinders. In the second case, the field has only an azimuthal component. Field distributions over the gap width in both cases are, correspondingly, of the form

$$\vec{H} = H_*(R/r)\vec{i}_r, \; \vec{H} = H_*(R/r)\vec{i}_\varphi \tag{9.38}$$

Here, H_* and R are constants.

The following identity holds true for fields (9.38)

$$(\vec{M}\nabla)\vec{H} = -[(\vec{M}\cdot\vec{H})_z\vec{i}_r + (\vec{M}\cdot\vec{H})_z\vec{i}_\varphi]/r$$

Taking account of this equality, the dynamic volume force for the azimuthal flow $\vec{v} = v(r, t)\vec{i}_\varphi$ may be presented as

$$\vec{F}' = (\partial\sigma'^{a}_{\varphi r}/\partial r + 2\sigma'^{a}_{\varphi r}/r)\vec{i}_\varphi \tag{9.39}$$

Here, $\sigma'^{a}_{\varphi r} = (1/2)(\vec{M}\times\vec{H})_z = \eta_r\,\partial_v/\partial r$. So, the projections of the equation of motion are of the form

$$\rho\frac{\partial v}{\partial t} = -\frac{1}{r^2}\frac{\partial}{\partial r}(r^2\sigma'_{\varphi r}) \; ; \; \frac{\partial p}{\partial r} = \frac{1}{r}\rho v_1^2 - \frac{1}{r^2}\mu_0 M_0 H_* R \tag{9.40}$$

Here, $\sigma'_{\varphi r} = (\eta + \eta_r)(\partial v/\partial r - v/r)$. Integration of equation (9.40) over the gap width for a steady flow gives the identity $R_1^2\sigma_{\varphi r}|_{r=R_1} - R_2\sigma_{\varphi r}|_{r=R_2} = 0$, where R_1 and R_2 are the radii of the internal and external cylinders, respectively. This equality, multiplied by 2π, represents the equality of rotational moments of the viscous forces acting upon the internal and external cylinders. There is no interaction of the moving fluid and the field source by means of force moments in this case. Therefore, the Couette flow in fields of cylindrical symmetry represents the most simple example of a dynamic interaction between the magnetic fluid and the field. This interaction can be explained by the magnetoviscous effect. As can be seen from (9.40), dynamics of a magnetic fluid, with the field on, is similar to that of a nonmagnetic fluid with effective viscosity $\eta_e = \eta + \eta_r$ which depends on the radial coordinate.

The developed flow profile is described by equation

$$(\eta + \eta_r)(v' - v/r) = \text{const}/r^2 \tag{9.41}$$

In "strong" fields, where the rotational viscosity coefficient is constant, equation (9.41) describes the normal Couette profile in the entire fluid volume. In "weak" fields $\eta_r = \eta_{r1} R_1^2/r^2$, where η_{r1} is a constant equal to the rotational viscosity near the internal cylinder. In this case, for the velocity profile to be determined from (9.44) we derive the equation

$$v' - v/r = \text{const}/(r^2 + SR_1^2)$$

where $S = \eta_{r1}/\eta$. Its solution, satisfying the boundary conditions $v|_{r=R_1} = \Omega_1 R_1$, $v|_{r=R_2} = \Omega_2 R_2$, is of the form

$$\frac{v}{r} = \Omega_1 + \Omega_2 - \Omega_1 \frac{\ln(1 + S) - \ln(1 + SR_1^2/r^2)}{\ln(1 + S) - \ln(1 + Sa^2)} \tag{9.42}$$

Here, $\alpha = R_1/R_2$. The rotational moment, acting per unit length of the cylinders, is

$$K = \frac{4\pi\eta_{r_1} R_1^2 (\Omega_2 - \Omega_1)}{\ln(1 + S) - \ln(1 + Sa^2)} \tag{9.43}$$

At $S \to 0$, (9.43) yields an expression for the rotational moment in a normal viscous fluid.

9.4 THE COUETTE FLOW IN THE FIELD OF A CYLINDRICAL MAGNETIC DIPOLE

Let us consider the flow of a magnetic fluid between rotating cylinders in the field of a cylindrical magnetic dipole. This example allows the effect of dynamic interaction for the Couette flow to be elucidated in greater detail than the flow in fields of azimuthal symmetry considered in the previous section.

The problem geometry is presented in Fig. 9.4. Cylinders of radii R_1 and R_2 are rotating at angular velocities Ω_1 and Ω_2. A cylindrical magnet of radius R

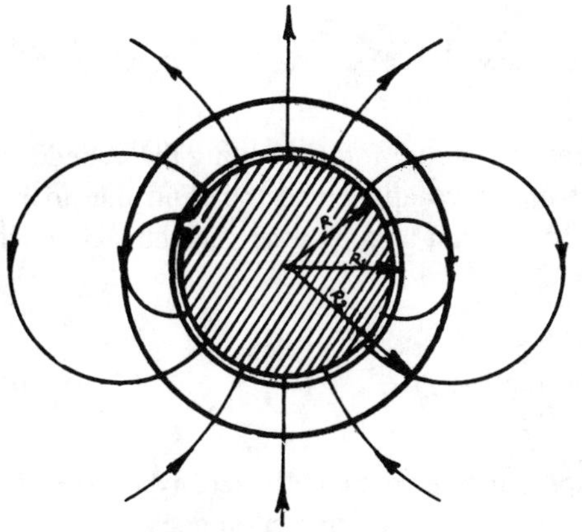

FIG. 9.4

and magnetization M_f, uniformly magnetized normal to the axis, is placed coaxially inside the internal cylinder and can rotate independently at velocity Ω. The field, induced by the cylindrical dipole, is of the form

$$\vec{H} = \frac{R^2}{r^2}\left(\frac{(\vec{M}_f \cdot \vec{r}\,)\,\vec{r}}{r^2} - \frac{\vec{M}_f}{2}\right) \tag{9.45}$$

In a noninductive approximation, the fields induced by the magnetic fluid may be ignored. Expression (9.45) yields the relations for the field intensity components and modulus in the cylindrical coordinate system

$$H_r = \frac{M_f R^2}{2r^2}\cos(\varphi - \Omega t); \; H_\varphi = \frac{M_f R^2}{2r^2}\sin(\varphi - \Omega t); \; H = \frac{M_f R^2}{2r^2} = H_* \frac{R^2}{r^2} \tag{9.46}$$

Using (9.46) gives the relation

$$(\vec{M} \cdot \nabla)\vec{H} = -(2/r)((\vec{M} \times \vec{H})_z \vec{i}_\varphi + (\vec{M} \cdot \vec{H})\,\vec{i}_r) \tag{9.47}$$

where $\vec{i}_\varphi$, $\vec{i}_r$ are the unit vectors of the cylindrical coordinate system. In view of the cylindrical symmetry of the problem, velocity distribution in the gap is sought in the form $\vec{v} = v(r)\,\vec{i}_\varphi$. For such a flow geometry, the distribution of viscous stresses is $\sigma'_{\varphi r} = \sigma'^{s}_{\varphi r}(r) + \sigma'^{a}_{\varphi r}(r)$ where

$$\sigma'^{s}_{\varphi r} = \eta(v' - v/r), \; \sigma'^{a}_{\varphi r} = \eta_r(v' - 3v/r + 2\Omega) \tag{9.48}$$

In order to elucidate the basic specific features of dynamic interaction between the magnetic fluid and the magnet, we shall use the integral equation of moment-of-momentum balance (6.49). For an annular fluid layer of a unit length, taking account of (9.47), it takes on the form

$$2\pi(R_2^2\sigma'_{\varphi r}\,|_{r=R_2} - R_1^2\sigma'_{\varphi r}\,|_{r=R_1}) + 4\pi\int_{R_1}^{R_2}\sigma^a_{\varphi r}\,r\,dr = 0 \qquad (9.49)$$

Here, the first term is the total moment of external forces rotating the cylinders; the second one, that of the moment of forces applied to the fluid due to its interaction with the field. Consequently, the unit length of the magnet is affected by the rotational moment

$$K = -4\pi\int_{R_1}^{R_2}\sigma^a_{\varphi r}\,r\,dr \qquad (9.50)$$

In the approximation linear with respect to the rotational viscosity coefficient η_r, K is calculated on the assumption that MF-field interaction does not distort the Couette profile (7.35). To simplify the calculation still further, we shall rely on the fact that the parameter ξ in the gap may vary within a limited range for which the dependence of rotational viscosity on the field (6.28) may be approximated by the square polynomial

$$\eta_r = \eta_r^{(0)} + \eta_r^{(1)}\xi + \eta_r^{(2)}\xi^2 \qquad (9.51)$$

where $\xi = \xi_1 R_1^2/r^2$, $\xi_1 = M_f R^2/(2R_1^2 H_T)$; $\eta_r^{(0)}$, $\eta_r^{(1)}$, $\eta_r^{(2)}$ are constants. As a result, we find from (9.50)

$$K = 4\pi\eta_r^{(0)}R_2^2\left[(\Omega_2 - \Omega_1)\left(1 + \frac{4a^2\ln a}{1-a^2}\right) + (\Omega_1 - \Omega)(1 - a^2)\right] +$$
$$+ 8\pi\eta_r^{(1)}R_1^2\left[(\Omega_1 - \Omega_2)\left(1 + \frac{\ln a}{1-a^2}\right) - (\Omega_1 - \Omega)\ln a\right] +$$
$$+ 4\pi\eta_r^{(2)}R_1^2\left[(\Omega_1 - \Omega_2)a^2 + (\Omega_1 - \Omega)(1 - a^2)\right] \qquad (9.52)$$

where $\alpha = R_1/R_2$. At $\xi \ll 1$, condition of "weak" fields is satisfied at any point in the gap and in (9.52) we may set $\eta_r^{(1)} = \eta_r^{(0)} = 0$, $\eta_r^{(2)} = \eta_{rs}/6$. At $\xi_1 a \gg 1$, the "strong" field approximation is valid and, hence, $\eta_r = \eta_r^{(2)} = 0$, $\eta_r^{(0)} = \eta_{rs}$. The case, when $\eta_r^{(0)} = \eta_r^{(2)} = 0$, $\eta_r^{(1)} \neq 0$ is conventionally referred to as a "moderate" field case. We shall only analyze these partial cases. Expression (9.52) can appropriately be compared with expression (7.39) for the rotational moment affecting the source of the rotating field, the flow geometry being similar.

If the cylinders are rotating at equal velocities $\Omega_1 = \Omega_2$ in (9.52) to make certain that in any of the above cases ferrohydrodynamic interaction always hampers the relative rotation of the field source and the fluid. As was shown in 9.1, the motion of a fluid as solid in "weak" fields was not distorted. Therefore, in this case the expression for K is exact.

In case of a shear flow, this interaction becomes more intricate. In Fig. 9.5, the relative rotational moment $K = K/(4\pi\eta_{r1}R_2^2\Omega_2)$ vs. parameter α is presented, when the magnet and the internal cylinder are at rest; for "weak" (curve 1), "moderate" (2), "strong" (3) and "uniform" (4) fields. It is seen that, in case of narrow gaps, the direction of dipole spinning is opposite to the cylinder rotation, no matter what the field intensity may be. In "strong" fields at $\alpha < 0.53$ and in "moderate" fields at $\alpha < 0.45$, $\vec{K}$ changes its sign if the direction of cylinder rotation is fixed. In "weak" fields, the sign of $\vec{K}$ is constant.

If it is the internal cylinder that is rotating, and the external cylinder and the magnet are at rest, the magnet will spin in the direction of cylinder rotation irrespective of the field intensity and the gap width.

The qualitative nature of interaction between the moving fluid and the field source may be interpreted using the following rule. This interaction always hinders relative rotation of the material element in the field direction. We shall illustrate the action of this mechanism for the flow under consideration. At any point in the gap the field rotational velocity is constant and equal to $\vec{\omega}_f = \vec{e}_x \partial \vec{e}/\partial t = -\Omega \vec{i}_z$, where $\vec{i}_z$ is the unit vector of the z-axis. The velocity of fluid element rotation relative to the field configuration, fixed at some instant of time, is the difference of the velocities of material element rotation relative to the laboratory reference system and of the field rotation relative to the material element in translational motion:

$$\vec{\omega}_* = 0.5\nabla \times \vec{v} - \vec{e} \times (\vec{v}\nabla)\vec{e} = [(2R_1^2/r^2 - 1)(\Omega_2 - \Omega_1)/(1 - a^2) - \Omega_1]\vec{i}_z = \omega_{f1}\vec{i}_z$$

It follows from this relation that at $\Omega_1 \neq 0$, $\Omega_2 = 0$ the sign of ω_{f1} for any α is determined by the sign of $-\Omega_1$. Therefore, the relative rotational velocity of particles and field $\omega = \omega_{f1} - \omega_f$ is specified by the difference $|-\Omega_1 + \Omega|$.

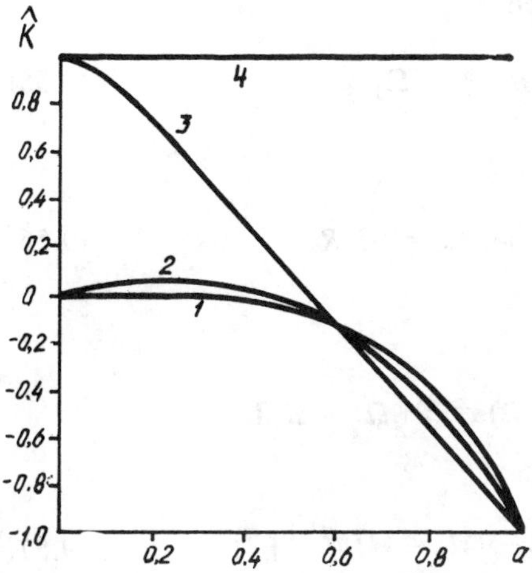

FIG. 9.5

Consequently, the magnet tends to turn towards the rotating cylinder, for in this case the difference $|-\Omega_1 + \Omega_2|$ will decrease. If $\Omega_2 \neq 0$, $\Omega_1 = 0$, for "narrow" gaps ($a > 0.71$) ω_{fl} has the same sign as Ω_2. In this case, the fluid-magnet interaction tends to decrease $|\Omega_2 + \Omega|$ and, hence, the magnet spins in the direction opposite to the cylinder. In a similar way, we may find the direction of ferrohydrodynamic forces in any other case if ω_{fl} for the whole of the fluid has a constant sign. For "wide" gaps ($\alpha < 0.71$) ω_{fl} changes its sign at $r = R_1\sqrt{2}$, with the result that K may reverse its sign as the gap width changes.

Consider the field effect on the rotational moments acting upon the internal and external cylinders. It is necessary to evaluate the field effect on velocity distribution in the gap. The projection of the moment-of-momentum balance equation (6.48) onto the axis z has the form

$$d(r^2\sigma'_{\varphi r})/dr + 2r\,\sigma^a_{\varphi r} = 0$$

With regard to (9.48) we obtain the equation for the velocity profile

$$(\eta + \eta_r)v'' + (\eta - 3\eta_r)\frac{1}{r}\left(v' - \frac{v}{r}\right) + \left(2\eta_r - \xi\frac{d\eta_r}{d\xi}\right)\frac{2}{r}\left(v' - \frac{3v}{r} + 2\Omega\right) = 0 \tag{9.53}$$

In the general case, integration of this equation is inconvenient because of a complex dependence of rotational viscosity on the independent variable. Let us restrict the consideration to "weak" and "strong" fields. For "strong" fields when $\eta_r = \eta_{rs} = \text{const}$

$$v'' - v'/r - n^2 v/r, + (n^2 - 1)\Omega/r = 0 \tag{9.54}$$

where

$$n^2 = (1 + 9\eta_{rs})/(\eta + \eta_{rs})$$

The general solution to this equation is

$$v = Ar^n + Br^{-n} + \Omega_1 r \tag{9.55}$$

Satisfy the boundary conditions

$$v|_{r=R_1} = \Omega_1 R_1;\; v|_{r=R_2} = \Omega_2 R_2 \tag{9.56}$$

to find

$$A = \frac{R_2^{1-n}}{1-a^2 n}\left[-(\Omega_1 - \Omega)a^{n+1} + \Omega_2 - \Omega_1\right];$$

$$B = \frac{R_1^{(1+n)}}{1-a^2 n}\left[\Omega_1 - \Omega - (\Omega_2 - \Omega)a^{n-1}\right] \tag{9.57}$$

The rotational moments, acting upon the internal and external cylinders, and the magnet, are, correspondingly, equal to

$$K_1 = 2\pi R_1^2 \sigma_{\varphi r}\,|_{r=R_1} = \frac{4\pi R_1^2 \eta}{1-a^3\, n}[\,S\,(\Omega-\Omega_1)\,(\frac{3-n}{1+n}a^{2n} + \frac{n+3}{n-1}) +$$

$$+ 1 + S\,(\Omega_2 - \Omega_1)\,n a^{n-1}\,] \qquad (9.58)$$

$$K_2 = -2\pi R^2 \sigma_{\varphi r}\,|_{r=R_2} = \frac{4\pi R_2^2 \eta}{1-a^{2n}}[\,(1+S)\,(\Omega_1 - \Omega)\,n a^{n+1} -$$

$$- S\,(\Omega_2 - \Omega_1)\,(\frac{3-n}{1+n} + \frac{n+3}{n-1}a^{2n})\,] \qquad (9.59)$$

$$K = \frac{4\pi R_2^2 \eta_r}{1-a^{2n}}\Big\{[\,(\Omega_2 - \Omega) - (\Omega_1 - \Omega)a^{n+1}\,]\,(1-a^{n+1})\,\frac{3-n}{1+n} -$$

$$-[\,(\Omega_2 - \Omega)a^{n-1} - \Omega_1 + \Omega\,]\,(1-a^{n-1})a^2\,\frac{n+3}{n-1}\Big\} \qquad (9.60)$$

In the above particular case, when the internal cylinder and the magnet are at rest, for the rotational moment acting upon the magnet (9.60) yields the expression

$$K = \frac{K}{4\pi\eta_r R_2^2 \Omega_2} = \frac{3-n}{1+n}\,\frac{1-a^{n+1}}{1-a^{2n}} + \frac{n+3}{n-1}\,\frac{a^{n+1}-a^{2n}}{1-a^{2n}} \qquad (9.61)$$

The dependence of the dimensionless rotational moment $\hat{K}$ on the parameter α for different values of n is presented in Fig. 9.6. It can be easily seen that sign

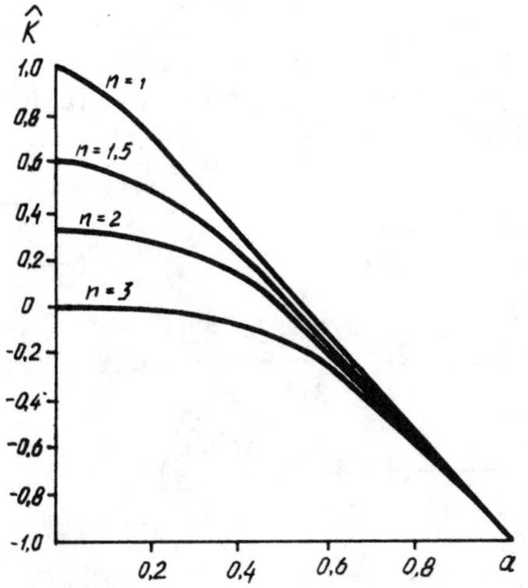

FIG. 9.6

reversal at K is most conveniently observed in the case of small $S(n \to 1)$ usually realized in practice. In the alternative limited case $S \gg 1$ $(n \to 3)$, which is not probably realized in practice, $K < 0$ and has a constant sign at any value of α.

Let us consider the case of "weak" fields. The rotational viscosity coefficient $\eta_r = \eta_{r1} R_1^4/r^4$. Here, $\eta_{r1} = \kappa H_1^2/4$ is the value of rotational viscosity near the internal cylinder. If $\eta_{r1} \ll \eta$, equation (9.53) includes a small parameter $S = \eta_{r1}/\eta$. Its solution will therefore be tried as S series expansion; $v = v^{(0)} + S v^{(1)}$. Here, $v^{(0)}$ is the Couette flow profile realized without a magnet. A correction for the distribution of viscous stresses in an approximation linear with respect to rotational viscosity, is related as

$$\sigma_{\varphi r} - \sigma_{\varphi r}^{(0)} = \eta_r (v^{(0)\prime} - v^{(0)}/r + v^{(1)\prime} - 3v^{(1)}/r + 2\Omega) \quad (9.62)$$

To estimate the first-order corrections for velocity, we obtain the equation

$$v^{(1)\prime\prime} + v^{(1)\prime}/r - v^{(1)}/r^2 = 8cR_1/r^2. \quad (9.63)$$

Here, $c = (\Omega_2 - \Omega_1)/(1 - \alpha^2)$. The solution of this equation, satisfying the boundary conditions $v^{(1)}|_{r=R_1} = v^{(1)}|_{R_2=r} = 0$, is of the form

$$v^{(1)} = (1/3)c[(1 + a^2)a^2 r - (1 + a^2 + a^4)R_1^2/r + R_1^6/r^5] \quad (9.64)$$

Relations (9.62) and (9.64) allow the calculation of the rotational moments K_1 and K_2 for "weak" fields. Let us cite expressions for the corrections for the rotational moments due to the field both for "weak" and "strong" fields at $S \ll 1$:

$$K_1 - K^{(0)} = \begin{cases} 4\pi R_1^2 \eta_{r1}[\Omega - \Omega_2 + (\Omega_2 - \Omega_1)\dfrac{a^4 + a^2 + 1}{3(1-a^2)}], & H_1 \ll H_T; \\ 4\pi R_1^2 \eta_{rs}[(\Omega_1 - \Omega)(1 + \dfrac{4\ln a}{1-a^2}) + \dfrac{\Omega_2 - \Omega_1}{1-a^2}(5 + 4\dfrac{1+a^2}{1-a^2}\ln a)], & H_2 \gg H_T \end{cases} \quad (9.65)$$

$$K_2 - K^{(0)} = \begin{cases} -4\pi R_1^2 \eta_{r1}[(\Omega - \Omega_1)a^2 + (\Omega_2 - \Omega)\dfrac{4a^4 - 2a^2 + 1}{3(1-a^2)}, & H_1 \ll H_T; \\ -4\pi R_2^2 \eta_{rs}[(1 + \dfrac{4a^2 \ln a}{1-a^2})\dfrac{\Omega_2 - \Omega}{1-a^2}(1 + 4a^2 + \dfrac{8a^2\ln a}{1-a^2})], & H_2 \gg H_T \end{cases} \quad (9.66)$$

Here H_2 is the field intensity near the external cylinder. The relations for "strong" fields ($H_2 \gg H_T$) are obtained from (9.58) and (9.59) in an approximation linear in the parameter S.

Let us discuss the qualitative features of the magnetodynamic effect $K - K^{(0)}$ for two particular cases: the impact of the rotating field on the fluid layer whose boundaries are at rest ($\Omega_1 = \Omega_2 = K^{(0)} = 0$, $\Omega = 0$); shear flow in a stationary field ($\Omega = \Omega_1 = 0$, $\Omega_2 \neq 0$, $K^{(0)} \neq 0$).

The magnetodynamic effect on the internal cylinder in a rotating field is positive at any values of the parameter both in "weak" and "strong" fields, i.e., the internal cylinder is spinning in the magnet direction. In case of a shear flow, the magnet increases the rotational moment (the magnetodynamic effect is positive) for "weak" fields at any α. In "strong" fields, the magnetodynamic effect changes the sign when the gap width is changed. In narrow gaps ($\alpha > 0.41$), it is positive; in wide gaps ($\alpha < 0.41$) it is negative.

The magnetodynamic effect on the external cylinder in the rotating "weak" field is negative at any α; it changes the sign in "strong" fields; it is negative in narrow gaps ($\alpha > 0.53$) and positive in wide gaps ($\alpha < 0.53$). In a shear flow, the magnet increases viscous friction on the external cylinder both in "strong" and "weak" fields.

The considered effects are attributed to the qualitative difference in symmetric and antisymmetric stresses. In the rotating field, symmetric stresses due to fluid motion in the magnet direction tend to turn both the external and internal cylinders in the magnet direction. The manifestation of this mechanism is weaker when the gap width is decreased and on a transition from "strong" to "weak" fields. Antisymmetric stresses strive to turn the internal cylinder in the magnet direction and the external one in an opposite direction. This mechanism is less sensitive to the varying gap width and field intensity. Thus, at any parameter values, both mechanisms strive to turn the internal cylinder in the magnet direction. No qualitative effects can be observed. The effects of antisymmetric and symmetric stresses on the external cylinder are opposite. In "weak" fields, however, symmetric stresses are not observed, and the direction of the rotational moment is determined by antisymmetric stresses. In "strong" fields the mechanism of symmetric stresses comes into force. It is not very distinct in narrow gaps and again, it is determined by antisymmetric stresses. As the gap widens, the fluid velocity grows and the rotational moment is reversed.

NOMENCLATURE

x, y, z—Cartesian coordinates
r, φ, z—cylindrical coordinates
r, φ, θ—spherical coordinates
t—time
$\vec{v}$—velocity
P—pressure
ρ—density
V—volume
T—temperature
ϑ—mean temperature deviation
$\vec{H}$—magnetic field intensity
$\vec{B}$—magnetic field induction
$\vec{M}$—magnetic moment of unit volume of the medium (magnetization)
$\vec{E}$—electric field intensity
M_s—saturation magnetization
I—current strength
$\vec{j}$—current density
σ_{ik}—stress tensor
σ^e_{ik}—Maxwellian field stress tensor
σ'_{ik}—viscous stress tensor
φ—volume concentration of solid phase
n—number of particles per unit volume
m—magnetic moment of a particle
μ_0—magnetic permeability of vacuum
$X_r = \partial M/\partial H$—differential magnetic susceptibility
$X_s = M/H$—integral magnetic susceptibility

$\mu = 1 + \chi$—relative magnetic permeability of a medium
σ—electric conductivity
α—surface tension coefficient
β_α—absolute temperature surface tension coefficient
β_ρ, β_M—relative temperature density and magnetization coefficients
$K = dM/dT$—absolute temperature magnetization coefficients
η, η_v, η_r—shear, volume and rotational viscosity coefficients
v—kinematic viscosity coefficient
λ—thermal conductivity
κ—thermal diffusivity
$\kappa_\parallel, \kappa_\perp$—dynamic magnetic susceptibilities
α_T—heat transfer coefficient
k—Boltzmann constant
g—gravitational acceleration
C—heat capacity
l—characteristic dimension
G—characteristic gradient of magnetic field intensity
γ—characteristic temperature gradient
$\vec{e}$—unit vector
k—wave vector
Λ—wavelength
ω—frequency
n—normal to the surface
R_1, R_2—main radii of surface curvature
δ_{ik}—Kronecker symbol
ϵ_{ikl}—Levi-Civita symbol

DIMENSIONLESS NUMBERS

R, Re $= vl/v$—Reynolds number
$\mathrm{Gr} = \beta_\rho g \gamma l^4/v^2$—Grashof number
$\mathrm{Pr} = v/\kappa$—Prandtl number
$\mathrm{Ra} = \mathrm{Gr} \cdot \mathrm{Pr}$—Rayleigh number
$\mathrm{Gr}_m = \mu_0 K G \gamma l^4/\rho v^2$—magnetic Grashof number
$\mathrm{Bo} = \rho g l^2/\alpha_2$—Bond number
$\mathrm{Bo}_m = \mu_0 M G l^2/a$—magnetic Bond number
$\mathrm{Si} = \mu_0 M^2/\sqrt{\rho g a}$—surface instability number
$\mathrm{Ma} = \beta_\alpha \gamma l^2/\eta\kappa$—Marangoni number
$\mathrm{We} = \rho l v^2/a$—Weber number
$\mathrm{Bi} = a_T l/\lambda_T$—Biot number
Nu—Nusselt number

INDICES

^{0}—values in mechanical equilibrium
*—constant mean critical quantities
′—small deviations from equilibrium values

NOTATION OF TRIGONOMETRIC FUNCTIONS

ch—cosh (hyperbolic cosine)
tg—tan (tangent)
th—tanh (hyperbolic tangent)
cth—coth (hyperbolic cotangent)
sh—sinb (hyperbolic sine)

REFERENCES

1. Sedov L. I. *Continuum Mechanics,* Vol. 1—M.: Nauka, 1970.—492p.
2. Landau L. D., Lifshits E. M. *Continuum Mechanics*.—M.: Gostekhizdat, 1953.—788p.
3. Landau L. D., Lifshits E. M. *Continuum Electrodynamics*.—M.: Nauka, 1982.—620p.
4. Vonsovsky S. V. *Magnatism*.—M.: Nauka, 1971.—1032p.
5. Gogosov V. V., Naletov V. A., Shaposhnikova G. A. Hydrodynamics of magnetizable fluids. In: *Science and Technology Results* (in Russian). VINITI, Ser. Liquid and Gas Mechanics, Vol. 16, 1981, p. 76–208.
6. Zahn M., Shenton K. E. *Magnetic Fluids Bibliography*.—IEEE Transactions on Magnetics, 1980, Vol. MAG-16, N2, p. 387–415.
7. Fertman V. Ye. *Magnetic Fluids: Convection and Heat Transfer*.—Minsk: Nauka i Tekhnika, 1978.—208p.
8. Luca E., Calugaru Gh., Badescu R., Cotae C., Badescu V. *Ferrofluidele si Aplicatiile Lov in Industrie*.—Bucuresti; Ed. Tehn., 1978.—336p.
9. Bluhm E. Ya., Mikhailov Yu. A., Ozols R. Ya. *Heat and Mass Transfer in Magnetic Field.* Riga: Zinatne, 1980.—354p.
10. Lykov A. V., Berkovsky B. M. *Convection and Heat Waves*.—Moscow: Energiya, 1974.—336p.
11. Berkovsky B. M., Vislovich A. N. *Effects of Volume Couples for Ferrofluids Moving in Magnetic Fields*. IVTAN Preprint N4-073—Moscow: 1981.—52p.
12. Rosensweig R. E. *Fluid Dynamics and Science of Magnetic Liquids*.—Advances in Electronics and Electron Physics, 1979, Vol. 48, p. 103–199.
13. Bibik Ye. Ye., Buzunov O. V. *Advances in Magnetic Fluid Production and Application*.—Moscow: TsNII 'Elektronika', 1979.—60p.
14. Shliomis M. I. *Magnetic Fluids*.—Uspekhi Fiz. Nauk, 1974, Vol. 112, N3, p. 427–458.
15. Bashtovoy V. G., Berkovsky B. M. *Thermomechanics of Magnetic Fluids*.—Magnetic Hydrodynamics, 1973, N3, p. 3–13.
16. Proceedings of the International Advanced Course and Workshop on Thermomechanics of Magnetic Fluids (Udine, Italy, 1977).—In: *Thermomechanics of Magnetic Fluids: Theory and Applications*. (B. Berkovsky, ed.).—Washington: Hemisphere Publ. Corp., 1978.—318p.

17. Proceeding of the Second International Conference on Magnetic Fluids (Florida, USA, 1980). —*IEEE Transactions on Magnetics,* 1980, Vol. MAG-16, N2.
18. Proceedings of the Third International Conference on Magnetic Fluids (Bangor, UK, 1983).—*Journal of Magnetism and Magnetic Materials,* 1983, Vol. 39.
19. Materials of the All-Union Seminar on Magnetizable Fluid Problems (Ivanovo, 1978).—Moscow: MGU, 1979.—78p.
20. Materials of the 2nd All-Union Magnetic Fluid School-Seminar (Ples, 1981).—Moscow: MGU, 1981—129p.
21. Materials of the 3rd All-Union Magnetic Fluid School-Seminar (Ples, 1983).—Moscow: MGU, 1983.—298p.
22. All-Union Symposium on Magnetic Fluid Hydrodynamics and Thermophysics (Jurmala, 1980). *Book of Abstracts.*—Salaspils: Institute of Physics, AN Latv. SSR, 1980.
23. Ukrainian Republican Meeting—Seminar on Thermophysical and Hydrodynamic Properties of Magnetic Fluids for New Industrial Technology and Refrigeration Engineering, *Book of Abstracts.*—Nikolaev: Oblpoligrafizdat. 1979.—39p.
24. Ferrohydrodynamic Problems in Shipbuilding. *Book of Abstracts,* All-Union Conf.—Nikolaev: Nikolaev Shipbuild. Inst. 1981.—92p.
25. 8th International Conference of MGD Energy Conversion, Vol. 5.—Moscow, 1983. p. 139–206.
26. 8th MGD Meeting in Riga (Riga, 1975). *Book of Abstracts,* Vol. 1, Riga: Zinatne, 1975.
27. 9th MGD Meeting in Riga (Riga, 1978). *Book of Abstracts,* Vol. 1.—Salaspils: Inst. Phys. Latv. SSR Acad. Sci., 1975.
28. 10th MGD Meeting in Riga (Riga, 1981). *Book of Abstracts,* Salaspils: Inst. Phys. Latv. SSR Acad. Sci., 1981.
29. 11th MGD Meeting in Riga (Jurmala, 1984). *Book of Abstracts,* Vol. 3: Salaspils: Inst. Phys. Latv. SSR Acad. Sci., 1984.
30. Convective and wave processes in ferromagnetic fluids.—Minsk, ITMO AN BSSR, 1975.—124p.
31. Convection and waves in fluids, Minsk.—ITMO AN BSSR, 1977.—159p.
32. Problems of magnetic fluid mechanics. Minsk: ITMO AN BSSR, 1981.—150p.
33. Magnetic fluids: scientific and applied studies. Minsk: ITMO AN BSSR, 1983.—142p.
34. Physical properties and hydrodynamics of disperse ferromagnetics.—Sverdlovsk: UNTs AN SSSR, 1977.—100p.
35. Physical properties of magnetic fluids.—Sverdlovsk: UNTs AN SSSR, 1983.—128p.
36. Neuringer J. L., Rosensweig R. E. *Ferrohydrodynamics.*—Phys. Fluids, 1964, Vol. 7, N18, p. 1927.
37. Tarapov E. I. *On hydrodynamics of Polarizable and Magnetizable Fluids.*—MG, 1972, N1, p. 3–11.
38. Moskowitz R., Rosensweig R. E. *Non-mechanical Torque Driven Flow of a Magnetic Fluid by an Electromagnetic Field.*—Appl. Phys. Lett., 1967, Vol. 11, 10, p. 301–303.
39. Kagan I. Ya., Rykov V. G., Yantovsky Ye. I. *On Dielectric Ferromagnetic Suspension Flow in a Rotating Magnetic Field.*—MG, 1973, N2, p. 135–138.
40. Berkovsky B. M., Vislovich A. N. Some convective transfer problems in ferromagnetic fluid. In: *5th All-Union Heat Mass Transfer Conf.,* Vol. 1, p. 2.—Minsk, ITMO AN BSSR, 1976, p. 251–260.
41. Suyazov V. M. *On Equations of Motion for Interpenetrating Electromagnetic Fluids.*—MG, 1971, N2, p. 12–20.
42. Berkovsky B. M., Vislovich A. N., Kashevsky B. E. *Magnetic Fluid as Continuum with Internal Degrees of Freedom.* Preprint ITMO AN BSSR, 1980, N4, p.41 B. M. Berkovsky, A. N. Vislovich and B. E. Kashevsky. *Magnetic Fluid as a Continuum with Internal Degrees of Freedom.* IEEE Transactions on Magnetics, 1980, Vol. Mag-16, N2, p. 329–342.
43. Tsebers A. O. *On Magnetization Models of Ferromagnetic Colloid Magnetization in a Hydrodynamic Flow.*—MG, 1975, 4, p. 37–44.
44. Shulman Z. P., Kordonsky V. I. *Magnetorheologic Effect.*—Minsk: Nauka i Tekhn., 1982.
45. Tables of Physical Quantities. Handbook. Moscow: Atomizdat, 1976.—1008p.

INDEX